# The Development of Garden Flowers

RICHARD GORER

EYRE AND SPOTTISWOODE LTD

*First published 1970*

*Printed in Great Britain for*
*Eyre & Spottiswoode (Publishers) Ltd.*
*11, New Fetter Lane, E.C.4.*
*Text by Northumberland Press Ltd, Gateshead*
*Plates by Fletcher & Son Ltd, Norwich*

SBN 413 27070 X

# Contents

# List of Illustrations

COLOUR PLATES



MONOCHROME PLATES

Where the source of an illustration has not been otherwise given, it has been taken from an issue of Curtis's *Botanical Magazine.*

## TEXT FIGURES



These are based on figures from *Plant Breeding* by A. L. Haagedoorn, published by Crosby Lockwood.

## ACKNOWLEDGEMENTS

Plates 23, 24, 25 and 26 are reproduced from *Pelargoniums* by Derek Clifford by kind permission of the author and the Blandford Press. Plates 16, 17, 18, 19, 20, 21, 22, 27, 28, 29, 30, 31, 32, 33, 34, 35, 36, 37, 38, 39, 40, 41, 42, 43, 44, 45, 46, 47, 48, 49, 50, 51, 52, 53, 54, 55, 56, 57, 58, 59, 60, 61, 62, 63, 64, 65, 66 and 67 are reproduced by kind permission of the British Museum (Natural History). Plate 7 is reproduced by kind permission of P. de Jager and Sons and C. G. van Tubergen. Plates 13, 14 and 15 are reproduced with permission of H.M. Stationery Office and the Director of the Royal Botanic Gardens, Kew. Plate 1 is reproduced from *Genetics of Garden Plants* by M. B. Crane and W. J. C. Lawrence (1947) by kind permission of Macmillan & Co. Ltd. Plates 2, 3, 4, 5, 8, 9, 10, 11, 12 and 68 are reproduced by kind permission of the Royal Horticultural Society.

## EPIGRAPH

'It would have proved most interesting had there been a reliable record kept by those who have raised the different forms of this most interesting plant, but horticulturists are not noted for keeping correct records of their doings, and consequently many important facts as to how new varieties of plants are developed are lost to the scientific world.'

*The Dahlia: Its History and Cultivation.*

# Preface

It came as rather a shock to me, when I started work on this book, to find that no one had previously gone into the subject. Since so many of the flowers that we grow in our gardens today or see for sale in the shops are the product of growers interfering with nature, it seems surprising that no one has traced their descent.

As I went further into the subject I understood the reason for this reluctance on the part of horticultural writers to tackle the subject. Either the material is not there at all and one is reduced to making more or less plausible guesses; or else the material is scattered in nineteenth-century gardening magazines and is difficult to find. I would be perfectly happy to spend a long time reading through these magazines, but it is a lengthy business, better undertaken in one's home than in libraries, so the number consulted has tended to depend rather on what was in my own or my friends' possession, than on the complete range.

It is always a problem when undertaking any research whether to continue for ever in the hope of achieving perfection or whether to do as much as possible and then make your results available to other workers. As I fear may be only too apparent, I have chosen the latter course. This book is apparently the first word on the subject, but it assuredly will not be the last. I have done my best to see that all glaring errors have been avoided, but no one is more conscious than I of the lacunae still left, in particular in the chapter on the Carnation. I can only plead that I have consulted every authority that either I or my friends could suggest.

The choice of what plants to include and which to ignore was less simple than I had imagined. My original scheme was to find the development of plants that varied so markedly from wild species as to make their identification difficult. There seemed to be no point in including every hybrid that had ever been introduced to the garden. So far the going was clear, but there then appeared a number of cases which had to be regarded as borderline.

In the case of some annual plants, including Clarkia and Godetia, I felt that the development had not been sufficient to warrant taking up space and the reader's patience, simply to note down a couple of species or so. Similarly having discussed a couple of greenhouse Primulas, I felt that it could be assumed that the others had been treated in the same way. It is in these marginal cases that it may be felt that I should have been more inclusive; on the other hand it may be thought that I have included too many plants that are not sufficiently distinctive to warrant the treatment I have given them. Should I have included Kniphofia and omitted Pentstemon? This is the sort of problem that I had to face and I can only hope to have arrived at a reasonably satisfying conclusion.

The help I have received from friends and colleagues has been very extensive. Above all I must thank Peter Hunt, who, whenever I was at a loss either produced the information I required or suggested a line of inquiry. Nine times out of ten this suggested line would bring the required information. Anthony Huxley was equally helpful in making suggestions and also in lending me books, enabling me to work in much more comfort even than that provided by the Lindley Library, whose librarian, however, Mr P. Stageman was also extremely helpful in making suggestions and who produced obscure volumes with amazing alacrity.

No one who has the privilege of Mrs Angela Marchant's acquaintance will be surprised to learn that there was no trouble she would not go to if it would be of assistance. Indeed the Iris world seems to attract most helpful people. Mr Maurice Peach and Mr Edwin Rundlett answered my queries about Remontant

Irises at great length, although both gentlemen were strangers to me.

Indeed if kindness and goodwill were all that were necessary to make a good book, this volume would be a masterpiece. Unfortunately other characters are also necessary.

Petham, Canterbury
January 1970

# Introduction

If it had not led itself to ambiguity, the best title for this book would have been Artificial Flowers, for we are going to discuss flowers that have been created by artifice. All the plants we shall be considering depart in some degree from the appearance of flowers growing in the wild. In some cases the difference may be due to selection of approved forms over a long period, while in others it may be due to the artificial creation of hybrids. Sometimes these hybrids have been continued for so many generations that they tend to behave like species and are propagated from seed. Tuberous Begonias, Calceolarias and Petunias are among many of which this can be said. In other cases selection has so modified the plant that it bears little relation to its wild ancestor. The Sweet Pea is perhaps the most conspicuous of these plants, but we can also mention such plants as Stock and *Iris kaempferi* and probably the Border Carnation.

Two much hybridized classes of plants have been omitted. These are Orchids and Rhododendrons, apart from the deciduous Azaleas. This is because the material is so very extensive and each subject would require a book on its own. The number of Rhododendron species in cultivation is extremely large and almost everyone has been used as a parent in hybridizing and there would be little point in just giving a list of these. Moreover the hybrids do not differ in any pronounced appearance from their parents. The same can be said to apply to orchids, with some reservations. The intergeneric hybrids of Orchids are, however, of interest and will be briefly referred to later.

It was only in the eighteenth century that the sexual nature of flowers was recognized and the process of pollination appreciated. Before then plant breeding had depended entirely on chance results followed by selection. Hybridization was carried out tentatively during the later eighteenth century, but it was not until the early nineteenth century that it was extensively practised. The impetus was chiefly due to the Hon. and Rev. William Herbert, Dean of Manchester, who wrote monographs on the Crocus and on the *Amaryllidaceae.* He published papers on hybridization, in which he argued that plants that were descended from a common ancestor would hybridize and were therefore species of a genus, but that one genus would not hybridize with another. It was owing to his work that *Azalea* ceased to be regarded as a separate genus and was combined with *Rhododendron.* In the course of his studies of the *Amaryllidaceae* he crossed species of *Gladiolus, Crinum, Hippeastrum* and *Narcissus* and the modern races descend from his beginnings. (The hybridization of *Crinum* was not persisted in.) Although, as we shall see (pp. 22ff.), it is possible to create intergeneric hybrids, these, with the exception of orchids, are usually sterile and have not proved of any value in the development of garden flowers. It is in the hybridization of species that the most spectacular results have been achieved.

It sometimes happens that a hybrid between two species, although an attractive plant, is sterile. Similarly plants with double flowers are generally sterile and even if either pollen or a stigma is present, will not necessarily transfer this doubleness to seedlings. If these plants are perennial, they can be propagated in other ways, by cuttings, layers, grafts, etc. In this book I have referred to this process as vegetative propagation. This is rather a clumsy phrase and inaccurate, as sowing seeds is also vegetative propagation. Asexual propagation would be more accurate, but again is not wholly correct as some plants can produce fertile seed asexually. Anyway for the sake of brevity I have decided to use the term vegetative propagation to describe methods of propagating plants by means other than sowing their seeds.

Hybrids are conventionally represented by the two species separated by X. In some of the earlier hybrids it is not clear which is the seed and which is the pollen parent, but I have tried to follow the convention in which the seed parent is the first named and the pollen parent the second. This convention of putting the seed parent first has only recently been established and some of the nineteenth-century writers were lax about it and one is not always certain even in more recent times.

I have also followed the practice of using the term cultivar (abbreviated as cv.) to indicate variation that has arisen as a result of cultivation. Botanists use the term *varietas* (abbreviated to var.) for deviations from the wild type found fairly consistently in nature. Thus the red form of the Primrose that is found more usually than the English type in Greece and Turkey is known as the var. *sibthorpii* of *Primula vulgaris*. On the other hand the double primroses are entirely a product of cultivation and have been given cultivar names in the vernacular. In the nineteenth century there was a strong tendency to give cultivars latin names, particularly if they were first cross hybrids. Thus we have Rhododendron 'Smithii' which nowadays would have to be given some name as 'Smith's Triumph'. This is decidedly a change for the better. It is highly confusing consulting Sweet's *Geraniaceae* and finding it almost impossible to differentiate between genuine species and subsequent hybrids, particularly when these may have had hybrids as each parent. Thus Sweet says that *Pelargonium barringtonii* is a cross of *P. crenaeflorum* and *P. involucratum*. These latter two were both hybrids, so that 'Barringtonii' is a second generation hybrid and none of them deserves a specific name.

CHAPTER I

# The bases of plant breeding

In this book we shall be endeavouring to find out how different races of garden flowers have been bred from wild species: some give the impression of being species in their own right; others by a process of selection have become so far removed in appearance from their wild ancestors that they seem to belong to a different genus.

When plants are brought into cultivation, a large number of things are liable to happen over the course of time. The cultivator will always select the most attractive or profitable forms from which to breed and so selection, on a scale that may be widely removed from Natural Selection, modifies the appearance and habit of the plant.

The essential basis of plant breeding is selection. The original gardeners must have started by selecting the plants they were going to grow in their gardens. Once a plant is chosen as being garden-worthy, it will probably be increased by sowing seeds. If any of the seedlings are improvements on the original plant, these will be selected for further development. Similarly one may get seminal variations: colour may be different or the leaves may be a different shape. These variations will be perpetuated if it is possible.

If the plant is ornamental but sterile for any reason, it will have to be propagated asexually by vegetative means; cuttings, layers, grafts etc. The same necessity will also occur with an $F^1$ hybrid, even though it may be fertile. Sometimes too there occur what are known as bud-sports. In these only a part of the plant shows a variation from the main form, which is due to a mutation in only a portion of the plant. Here, again, vegetative propagation is the best method of perpetuating the variation.

A population that is derived asexually from a single plant is known as a clone. Failing any mutation, all members of the clone will be identical in make-up to the ancestral plant. This is, of course, very convenient, but should the clone be susceptible to some virus or fungal disease it may well be lost. It is generally believed that clones in time become degenerate, but this is not necessarily so. The Dutch Yellow Crocus has been known since the sixteenth century and shows no sign of degeneration. The Double Daffodil, 'Van Sion' dates from the seventeenth century. Many of the older roses have survived from before the sixteenth century, but in general it is the bulbous and tuberous plants that have survived the longest. Plants that have survived through cuttings or grafting do not show so long a history in Europe. This, however, may be because they have been superseded by better cultivars. There are many Chinese plants that appear to have remained for centuries in cultivation. Against this there are cases where the stock has weakened and eventually become extinct. Farrer's var. *major* of *Edraianthus pumilio* is rapidly disappearing and many hybrid lilies have proved far too susceptible to virus. The first *Lilium auratum-speciosum* hybrid, *L.* X *parkmannii,* was lost through virus. It is obviously preferable to have plants that can be reproduced seminally, but it is not always possible.

In the course of time, it is probable that seminal variation may be considerable and these varieties can be interbred; the resultant plants are known as crosses. It may also be possible to interbreed different species of the same genus; this is hybridization and gives rise to hybrids. On some occasions it may be possible to hybridize different genera, although, as we have seen, the late Dean Herbert believed that this is *a priori* impossible and that if such hybrids do arise the plants have been wrongly consigned to different genera. Nowadays we have a number of intergeneric hybrids, but Herbert's theory is not necessarily disproved by them. Many of the intergeneric hybrids such as X *Fatshedera*

(*Fatsia japonica* X *Hedera hibernica*), X *Phyllothamnus* (*Phyllodoce empetriformis* (?) X *Rhodothamnus chamaecistus*) are between genera that are widely separated geographically and may well represent different development from a common ancestor. The same cannot be said of the hybrid between *Philesia magellanica* and *Lapageria rosea*, nor of the crosses between *Heuchera* and *Tiarella*. Here the plants are geographical neighbours and, if the late Dean's theory is correct, botanists are therefore wrong in assigning them to separate genera.

Among the *Gesneriaceae* there are some members – notably *Isoloma, Sinningia, Gesneria, Achimenes,* and *Naegelia* – which all seem to cross with great ease and hybrid races such as X *Paschia* (between *Isoloma* and *Sinningia*) have been raised. However the status of all these genera is by no means certain and the family is continually subject to taxonomic revision, so it may be safer to regard these intergeneric hybrids as slightly doubtful. There is no question of the hybridity, but the genera may not yet be accurately defined.

The case of Orchids is perhaps slightly different. The various plants do appear generically distinct in most cases, but it is significant that the intergeneric hybrids are almost all between genera from the same continent. I know of only one instance where, for example, a South American genus of orchid has been crossed with an Asiatic one. There are numerous intergeneric crosses between South American orchids and, more recently, between Asiatic ones. The only record of a cross between Asiatic and American orchids seems to be two cases of the South American *Phragmipedium* being crossed with the Asiatic *Paphiopedilum* (the orchid always known incorrectly as *Cypripedium*, while *Phragmipedium* is often known as *Selenipedium*). As, however, these orchids are patently derived from a common ancestor (probably *Cypripedium* which is found in America, Europe and Asia), this scarcely disproves Herbert's theory. Orchids still seem to be in an active state of evolution and it would be possible to argue that many of the genera are very artificial constructs. However that may be, there are cases where several genera have been interbred in the orchid family; for example X *Potinara* contains the blood of *Sophronitis, Rhyncolalelia* (*Brassavola*), *Laelia* and *Cattleya.*

In general intergeneric hybrids are uncommon, and, except in the case of orchids, sterile. On the other hand interspecific hybrids are frequent and so are subsequent crosses between these hybrids. Indeed it is through the crossing of hybrids that new races often arise, as the rest of the book will show. However, in the case of species that are naturally variable, a number of plants can be raised that are sufficiently distinct from each other to warrant the attribution of race and cultivar names, without any other species being concerned. Such plants as the Sweet Pea and the Snapdragon may be cited as examples.

During the nineteenth century plant breeding became more scientific and it was recognized that the selection of the best forms for breeding was important, but, although Mendel's famous paper on Peas was published in 1865, the working of inheritance was not understood and crosses were generally stopped at the first generation. Writing in 1868 about raising Zonal Pelargoniums Shirley Hibberd says dogmatically that all unsatisfactory crosses should be destroyed. Nowadays we would realize that the second generation might well produce the plant we are looking for and would self the first generation, however unsatisfactory it might be. Let us try to understand why this is so.

The subject is extremely complicated and many large volumes are filled with the theories and explanations. What follows is drastically simplified. As I am no geneticist myself I have relied on the writings of others, notably the *Genetics of Garden Plants* (by Crane and Lawrence, London, 1947), *Genetical Principles and Plant Breeding* (by W. Williams, Oxford, 1964) and *Plant Breeding* (A. L. Hagedoorn, London, 1950). It is perhaps unnecessary to add that any errors in what follows are my own and should not be attributed to the authorities consulted.

Like all living creatures, plants are composed of cells. The operative part of the cell, embedded in a substance called cytoplasm, is the nucleus, in which are a number of elongated bodies, known as chromosomes. These chromosomes in normal plants consist of an even number. Plant growth is caused by cell division, so that where there was previously one cell there are now two. During cell division the chromosomes behave in a very characteristic way. The process is known as mitosis. First each chromosome reproduces itself, dividing longitudinally, so that if there were originally 6 chromosomes in the cell, there are now 12 half-size chromosomes, known as chromatids. These chromatids contract and form a rod-like structure in the centre of the nucleus. They then, as though mutually repulsed, move to opposite ends of the cell and a new cell wall is formed between the two daughter nuclei. There is a resting stage, during which the chromatids enlarge to the size of the original chromosomes and the process is then repeated. In this way the cell retains the original number of chromosomes in each cell. Moreover these have all come from the original pair of chromosomes in the fertilized ovule, so that the plant is derived from one original cell. It is, however, possible on occasions for something to go wrong. If, for some reason, the new cell wall does not form, we are left with a cell with twice as many chromosomes as had been in the former cells. If, after this one failure, the normal cell division is resumed, we will be faced with a part of the plant that has twice as many chromosomes as the main body of the plant. It is likely that the cells containing more chromosomes will be larger than the cells that contain only half as many and the plant may therefore be desirable if, for instance, it has larger flowers. A normal plant is known as a diploid, each cell containing twice the number of chromosomes that are basic to that plant. The basic number is known as the haploid number. If, in the case we have been positing, the cell contains four times the haploid number, it is known as a tetraploid. Plants with cells containing more than two sets of chromosomes are known as polyploids. As a general rule polyploids are all multiples of the haploid number, but occasionally it may happen that an additional chromosome or so may be added or subtracted.

It is not very often that part of a plant becomes polyploid in the way that we have been describing, but this can sometimes be induced by shock. If you cut off the growing point of a tomato plant, it is quite possible that one of the resultant side-shoots will be tetraploid. This can be recognized by the larger leaves. Nowadays polyploidy can be induced by the use of colchicine and this is often employed by plant breeders.

Most of the higher plants, with which this book is concerned, reproduce themselves by fertile seed. Seed is produced as a result of fertilization of the ovule by pollen; but if the ovule and pollen cells were formed in the way we have been discussing, the seeds resulting from their union would contain twice the number of chromosomes of the original plant. When the ovules and pollen are formed, therefore, a different process, known as meiosis or reduction division, comes into play.

During the first stage of *mitosis*, a cell's chromosomes double in number; during the first stage of *meiosis*, the number of chromosomes remains the same, although their character may alter.

It will be remembered that a normal diploid cell contains two sets of chromosomes. At meiosis, these form themselves into two necklace-like strands, which are attracted to each other. The two strands come together and merge, each single chromosome uniting with its equivalent in the other strand. After merging, the pairs separate again into single chromosomes, and from this point follow the same process as in mitosis, the reconstituted chromosomes acting as do chromatids. At the end of meiosis we therefore have four cells, each containing only half the normal number of chromosomes. These haploid cells develop to form germ cells in pollen or ovules. Fertilization will recombine the two sets of chromosomes and form the basis for a new plant with the normal diploid number of chromosomes.

To the plant breeder, the importance and interest of germ cell formation lies in the first stage of meiosis. As a result of the merging process, all the chromosomes in a cell are reconstituted, some in a straightforward manner (Fig. 1) and some by a phenomenon known as 'crossing-over' (Figs. 2 and 3) in which each new chromosome is made up partly from one of the original pair, partly from its twin. Where the crossing-over is regular (Fig. 2) the resultant chromosomes will be identical to the original pair. But where the crossing-over is irregular

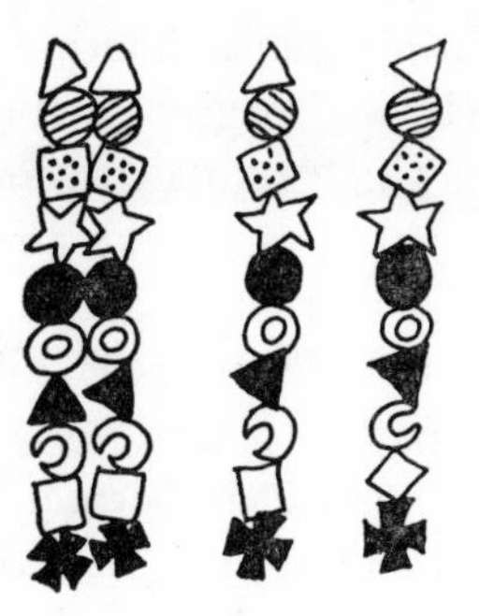

Fig. 1.
Diagrammatic representation of a chromosome pair during meiosis, showing the redistribution of genes after the separation stage. The different shapes represent different genes, and it will be seen that in the two new chromosomes these are identical in number and pattern to those in the two original chromosomes.

Fig. 2.
Regular crossing-over of two members of a chromosome pair during meiosis, showing that here also the arrangement and number of genes remains identical.

Fig. 3.
Irregular cross-over, as a result of which one chromosome in the pair has gained a gene, and the other chromosome has lost it.

(Fig. 3) the result will be two chromosomes differing from the original pair. It is this fact which, we imagine, is the cause of seminal variation, since the genes, which regulate various growth factors, are situated on the chromosomes and an irregular crossing-over will cause a new distribution of genes. Seedlings arising from the subsequent germ cells may be different in vigour or appearance from the parent plants.

The preliminary pairing in meiosis is important, and will usually only take place if the chromosomes are identical or nearly so. As we saw, each cell of a normal plant has its chromosomes in pairs. If we create an interspecific hybrid, we may find that it is sterile, either because exact pairing at meiosis is impossible or because, though pairing takes place fairly completely, subsequent segregation during division is random, producing infertile pollen or ovule cells. However, tetraploid forms of interspecific hybrids are fertile because every chromosome is provided with an identical partner and so one complete set from either parent is present in every gamete; pairing can therefore always take place at the beginning of meiosis.

The creation, by use of colchicine of tetraploid forms of sterile interspecific hybrids is recognized as a way of obtaining fertility. Such plants are sometimes termed amphi-diploid hybrids. If the plants have considerable affinity, as in the case of *Petunia axillaris* and *P. nyctaginiflora,* the diploid hybrid will be fertile and no further action is necessary to breed a hybrid race, than to sow the seeds of the original cross. There are many other genera and species where the hybrid proves to be fertile, but in the event of it proving sterile, it will often be found that doubling the chromosomes will give fertility. It sometimes happens that species will not cross in spite of a similar chromosome count and here again, if both species are made tetraploid, it is possible that a hybrid can be raised, but this is by no means certain. Sometimes the inability of plants to hybridize may be due to purely physical causes. If pollen from a short-styled plant is put on to the stigma of a plant with a much longer style, the pollen tube may be unable to penetrate the ovule.

Sometimes during meiosis, reduction division may fail so that both pollen and ovules are diploid and not haploid. If the plant is self-fertilized we should get a tetraploid, but if the diploid ovule is fertilized by haploid pollen we get a triploid: the cells contain three times the haploid number. For instance, in the case of a Primula with a haploid count of 9, the triploid plant would have 27 chromosomes. Now these cannot pair exactly at the first stage of meiosis and so the plant is liable either to lack pollen or ovules at all, or if it has them they are liable to be useless. As a general rule plants with an odd multiple of their haploid count, 3, 5 or 7 are likely to be sterile. There are a few rather unusual exceptions. The most notable of these is the Dog Rose, *Rosa canina*, agg., which has 35 somatic chromosomes.*

Pairing is sometimes irregular with polyploids that one would expect to be fertile. Microscopic examination will show that the chromosomes may have separated into groups of three and one, rather than in the two pairs or group of four that might be expected. This seems to occur far more frequently in autopolyploids. Allopolyploids, which are of hybrid origin, usually pair regularly.

The units of reproduction in the hereditary material are known as genes. These are mainly, if not entirely, situated on the chromosomes and we know very little about them. They appear to be complex chemical substances that have a power of self-reproduction. They are so minute that they have never been seen, but it is due to their action that plants reproduce themselves. (Not only plants, of course: the more or less identical reproduction of all living things is due to their action.) The number of genes in any cell is considerable and not all chromosomes seem to bear the same number. One chromosome may carry a large number, while another may have only relatively few. It is generally believed that each gene is responsible for some feature of the plant. It is rather like feeding information into a computer: each gene has its own instruction. Now we have seen that during the process of crossing over some chromosomes may be reconstituted so that part of one joins with

* See chapter V, p. 90.

a corresponding part of another. While this is happening some of the genes on the chromosome may get redistributed and we may get a chromosome without a particular gene, while another may receive double the normal amount. This is one of the ways that variation arises. Genes that are located on the same chromosome are said to be coupled or linked and though it is sometimes possible to separate the genes that are far removed from each other on the chromosome, if they are very near to each other it is extremely difficult.

You may remember that each cell has its chromosomes in pairs and therefore it has its genes in pairs. Sometimes it would appear as if the amount of the gene tends to vary, with a corresponding effect on the plant. These smaller quantities are termed recessive genes, while the larger amounts supply dominance. If we cross a plant which has a visible expression of a recessive gene with one that lacks this, the resultant plant will resemble visibly the dominant form. Genetically, however, it will be different. This is perhaps best illustrated by taking an actual case. Botanically the Peach and the Nectarine belong to the same species, but can be distinguished by the peach having a downy skin, while the nectarine has a smooth one. If we pollinate a nectarine with a peach, or *vice versa,* the resultant plants will all resemble a peach and have downy skins. We say that the gene *A* for hairy skins is dominant over the gene *a* giving a smooth skin. If, however, we self-fertilize this offspring of the peach and nectarine, we find in the next generation we get one nectarine to every three peaches. The explanation is simplest in diagrammatic form.

Peach *AA* Nectarine *aa*

Gametes (pollen or ovules)

*A* *a*

First generation *Aa*

Gametes *A* or *a*

Possible combinations *AA*, *Aa*, *aA*, *aa*.

As *A* is dominant, the *AA*, *aA* and *Aa* plants all appear as peaches. The *AA* plants will breed true and are referred to as homozygous and so will the *aa* plants, but the *Aa* and *aA* plants are heterozygous and will continue to produce offspring as the $F^1$ hybrid did.

We have referred to a quantitative difference in the make-up of the genes, but this is by no means a universally accepted theory. It is more general to assume that all genes have a recessive counterpart, which occasionally emerges as the result of a mutation. A mutation means just what it says, a change, and the result of a mutation is that the function of the gene changes. Whether it is due to a quantitative difference or a molecular rearrangement is by no means certain. Some genes appear to have more than one recessive character and are known as multiple allelomorphs, but this does not occur with all that frequency. A quantitative explanation certainly would seem a more satisfactory explanation here.

The case of the peach we have just discussed is that of a plant with only one genetic factor distinguishing it from another. The typical 1 : 2 : 1 Mendelian ratio is masked by reason of the dominance of the gene *A* over *a*. If neither gene is dominant over the other, the Mendelian result is much clearer and is memorably illustrated in the limerick about the young girl of Osaki, who spent a week-end with a darkie. The result of their sins was quadruplets (not twins); one yellow, one black and two khaki.

Generally in plant breeding you are dealing with a large number of different genes and the ratio necessary to get the emergence of more than one recessive gene becomes larger according to the number of recessive genes that have to be extracted. Thus the ratio is 3 : 1 with one gene, 15 : 1 with 2 and 63 : 1 with three recessive genes.

We tend to think of each gene as being responsible for a definite character in the plant, but this is not always so. Sometimes more than one gene is necessary to produce an effect. Thus in the Sweet Pea there are two genes that are necessary to produce anthocyanin, the purple colouring in the petals. These have been isolated as C and R. Plants that have only C will have white flowers and so will plants that have only R. If, however, we cross the white C plant with the white R plant, all the offspring will have purple flowers, which is an unexpected result of the crossing of two white flowers. Of course C plants crossed with C plants or R plants crossed with R plants will continue to give white flowers. Where it is found that two or more genes are necessary to give an effect, the genes are known as complementary.

In their classic work on *Primula sinensis* De Winton and Haldane* described the action and interaction of 25 genes, together with their mutants and localized a number of them on their chromosomes. Thus on one chromosome there are five genes. *S* causes a short style and long anthers, while the allele gives a long style and short anthers.† *B* causes a bluish colour and increases the yellow. Its recessive counterpart will give red flowers. *X* is responsible for the creation of stamens and styles, while its mutant gives sterile double flowers. *G* suppresses anthocyanin in the flower centre, while its allele gives flowers with a dark centre and a red stigma. *L* suppresses anthocyanin in the leaves and stems, while its mutant gives dark purple stems and leaves.

All these genes are present on the one chromosome and it was calculated that to separate the gene *G* from *L* during crossing-over would occur only once in 6,500 plants, in the second generation. This is approximately what happened and at length a plant was raised with dark flowers and purple leaves and stems.

It will be noted that the result of the mutation of any gene could be stated. As a general rule these mutations, when they occur, are recessive and only segregate out in the second generation, unless they can be selfed. Some of the mutants look more like absences. The mutant of *X* which gives sterile double flowers looks as though the *X* gene, which causes the formation of

* D. de Winton and J. B. S. Haldane, 'The genetics of *Primula sinensis*', *Journal of Genetics*, 1933.

† One tends to raise one's eyebrows slightly over the gene *S*. Most Primulas normally have pin and thrum flowers on separate plants and it could be assumed that *S* will cause one effect or the other. It seems improbable that heterostyly is due to a mutation.

stamens etc., is absent altogether and the mutants of the *G* and L alleles look as though the difference may be quantitative. Indeed if we look at the list of genes and their functions that De Winton and Haldane give for *Primula sinensis* and the similar list that Punnett gives for the Sweet Pea,* one is struck with the fact that in most cases the function of the mutant is the complete opposite of that of the normal gene. In such cases the presence or absence theory of genes does seem to make sense; more sense at any rate than the idea that every gene carries its mirror image *in posse*. If the gene for forming stamens etc., is absent, while that for the formation of petals is present, one would expect sterile double flowers.

Such an explanation will not be valid for the multiple allelomorphs. These genes vary in their end results. Thus the size of the eye in the flower of *P. sinensis* is controlled by the allele *A*. In its most intense form *A'*, it suppresses the plastids altogether, but in its forms *A* and *a* it restricts them to a greater or lesser degree. Flowers with *A* will have a small yellow eye, while those with *a* will have a large one. Here it looks as though the difference is quantitative. *A'* looks as though it has a larger dose of the plastid-suppressing substance than either of the other two alleles.

If I have succeeded in giving the impression that a plant is an extremely complicated being in which many genes combine to give what we may term the typical form, it will be appreciated that plant breeding is by no means simple and that, in spite of scientific discoveries, there will remain a good many imponderables. However, one fact has already been established that is of value to the plant breeder. In any hybrid population the recessive plants will be in a minority and they can be relied upon to breed true. Thus if we wish to get a recessive flower colour, it can be recognized by the fact that it is the colour least frequently seen in the offspring of the second generation. If the plant can be self-fertilized, or if there are more than one of the same colour, it will breed true. The recessive genes are always

* Proc Roy. Soc. 102, 1927.

the easiest to fix, once they have been brought out.

It would be a mistake to think of some genes as invariably recessive and some as always dominant. In every case the dominance is relative and it is possible that what is dominant in one set of plants might be recessive in another. It is, however, a convenient shorthand to talk of dominant and recessive characters. The technical terms epistatic and hypostatic are more accurate, but are less popular.

Once selection has gone on for some time, one may often get very spectacular results from crossing two cultivars. These first filial or $F^1$ crosses, as they are called, exhibit marked hybrid vigour (heterosis) and have of late become very popular with vegetable growers. Some less spectacular results have been obtained with such plants as Petunias. Since the production of these crosses needs hand pollination and, occasionally if the plants are self-fertile, emasculation, the production of $F^1$ crosses is expensive. The explanation, for what it is worth, of this hybrid vigour, is that both parents contribute genes that are favourable for growth and development and that these genes are dominant, so that the $F^1$ cross has two or more genes that are favourable to growth and development.

It seems to be generally assumed that all the genes are located on the chromosomes, but there seems reason to doubt whether this is in fact so. If we cross two plants we may get quite different results according to which plant is used as the pollen parent and which as the seed parent. Since the chromosomes should be the same in each case, one would expect similar results, but this does not happen; a fact of which the old plant breeders were well aware. Shirley Hibberd, a first-class and much underrated horticultural writer, wrote in 1868: 'The seed parent usually has most influence on the form of the flower and leaves; and the pollen parent usually has the most influence on the colour of the flowers and the leaves. To a certain extent however, the characters of both parents are usually blended in the offspring.' In the 1880s Dr Stuart, a famous raiser of Violas wrote: 'In hybridizing wild species of *Viola* with pansies, it is necessary that the pollen be taken from the cultivated species of Pansy and dusted over the pistil: that is that the wild species should be the mother. Pollen taken from *V. cornuta,* for instance, will, if put on the common Garden Pansy, only give seed which will produce Bedding Pansies, not the sturdy tufted-rooted dwarf strain.' Crane and Lawrence* quote experiments by Passmore with Vegetable Marrows. She made crosses between large and small-seeded cultivars and found that when the large-seeded type was used as the seed parent the seeds were larger than in the reciprocal cross. There are quite a number of examples of this phenomenon which is termed extra-nuclear inheritance and which appears to be connected with the cytoplasm of the mother plant.

Less common, but still to be observed with some regularity, is the influence that the pollen may have on the plant. This is usually quantitative and does not seem to be inherited. Thus if a sweet almond is pollinated by a bitter almond, the resulting almonds will taste bitter. Dates when pollinated with one cv. will give small fruits that ripen earlier than the majority and crosses between cvs. of the Sweet Chestnut will cause the development of nuts larger than those of either of the parents. This latter phenomenon could be due to heterosis.

The importance of the maternal cytoplasm seems until recently to have been somewhat overlooked by writers on the subject, but it certainly looks as though some genes must be located away from the chromosomes and are found in the cytoplasm. Such genes, one would imagine, must be very stable and unlikely to show mutation. There is, so far as we know, no crossing-over in the cytoplasm. Thus when the nineteenth-century gardeners wanted to combine the colour of *Rhododendron arboreum* with the hardiness of *R. catawbiense* they were careful to use the hardy plant as the seed parent, assuming that hardiness was a maternal quality. The reciprocal cross might have proved equally hardy, but they did not think so and any gardener will appreciate that years of empirical observation are worth more than a few scientific experiments.

We saw earlier that it is comparatively easy to isolate a single

* In *The Genetics of Garden Plants* (London, Macmillan, 1947).

recessive gene in a hybrid strain; but some of the intermediate forms may be desirable, but difficult to 'fix' so that a true breeding strain may be produced. Sometimes breeding further generations will enable us to do this, while at other times back-crossing to one of the original parents may be the answer. This is often the case with disease resistance. This is generally associated with a dominant gene, but it may occur in a plant that is undesirable in other features. If we cross this with a desirable form, the $F^1$ will probably have many of the qualities we require combined with disease resistance. If we now back-cross to the desirable plant we should get nearer our goal and a second back-cross may well give us the plant we require. For safety's sake it might be advisable to grow on the $F^2$ generation and back-cross this as well. There is a danger when hybridization has gone on for too long a time that the number of potentially useful genes may have become diminished and there is often much to be said for re-introducing fresh importations of the original species parents into the strain. This will generally provide additional vigour and may also show increased variability.

The plants of polyploids are usually, though by no means invariably, larger than diploids for the simple reason that the individual cells have to be larger to contain the increased number of chromosomes. The polyploid plant therefore is 'larger in all its parts', to quote a good botanical cliché and as size in itself is a quality that attracts many gardeners, the polyploid is usually selected for propagation. The best method of inducing polyploidy is by treatment with colchicine. Seedlings or the growing tips are treated with a very dilute solution: usually something between 0.1 to 0.4% is the maximum dose, although 1% is recommended for some hard-wooded plants. When it is applied to the growing tips of plants, the colchicine is sometimes incorporated in lanolin and the treated grease is smeared on to the tip. It is essential that colchicine should only be applied when the plant is making rapid growth and this must be encouraged by greenhouse treatment or something similar, so that the conditions are suitable for optimum growth and so as to ensure that there is no check to the speed of cell-division. The action of the colchicine is, apparently, that it inhibits the separation of the split chromosomes during meiosis. You may recollect that after splitting, they separate to opposite ends of the cell, while the cell wall forms between them. Colchicine appears to inhibit the separation of the chromatids so that no cell wall is formed and cells are produced with double the basic number of chromosomes. When, later, normal cell-division is resumed, part of the plant will be tetraploid or possibly the whole plant will be. In any case the diploid portion will be visible and can be removed. The process is somewhat critical and losses among treated seedlings are quite high.

With the large number of genes that every plant has and the consequent possibilities of variation, it is not surprising that the great majority of man-made flowers are to be found either among plants that can be brought rapidly into flower from seed, or else among plants that have a long history of cultivation in gardens behind them. It is generally the annuals and biennials that have been studied in depth by the geneticists and in such cases future breeding is a little less empiric than is the case in most plant breeding, even today.

In earlier days, when many of our best plants were being raised, plant breeding was entirely the result of selection. This selection could take place either within the offspring of one species or within the offspring of a hybrid.

As an example of the first technique, we might consider the development of the Shirley Poppies. It was in 1880 that the Rev. W. Wilks, then vicar of Shirley (later he became Secretary to the R.H.S.) noticed in a cornfield a scarlet poppy that had a white edge to its petals. He marked the plant and collected the seed. Most of the offspring were the ordinary scarlet poppy and these were all destroyed; any that showed variation were kept and the seeds sown. This process was continued for many years until finally a strain was built up that would give plants ranging in colour from white, through various shades of pink to crimson. In fact what the Rev. Wilks did was to select plants that showed the influence of recessive genes in the formation of petal colour. Nowadays this could probably be done in less time, by controlling the pollination. It is remarkable how many plant breeders have used rigorous selection, but made no attempt to control the pollination. Russell with his Lupins and Luther Burbank with a number of different plants appear to have relied entirely on random pollination to be followed by selection in the resulting generations. Since in each case they obtained results, it is not for us to sneer at such methods, but it does appear to be rather a lengthy process.

Somewhat different was the development of the Zonal Pelargonium. This is discussed at some length in Chapter IX but here it is enough to say that it was the result of a hybrid between the large-flowered *Pelargonium inquinans* and a species with a purple horseshoe-shaped mark on the leaf and rather small irregular flowers. The resultant hybrid showed that both the desirable features were dominant. The plant had large flowers and the ornamental leaf. Moreover the hybrid was fertile and further generations were bred. Nowadays one would breed further generations even if the original cross had not produced a satisfactory plant. It is this, rather than any knowledge of genes, that distinguishes modern plant breeding from that of the last century. It is only when plants are raised in large numbers that any amount of mutations can fairly confidently be expected. It is for this reason that it is among the annuals, and plants that will flower the first or second year after sowing, that the greatest variability is to be observed although something similar is to be found in plants that have been in cultivation for a long time. Thus many of the plants of Chinese gardening, as we shall see in Chapter III, which have now been in cultivation for more than 2,000 years, will show very considerable variation, even in plants like Camellias, which take many years to flower from seed. Generally it will be found that shrubs and trees show far fewer evidences of mutation than do plants that are raised in large numbers. Of course, it must be remembered that environment can produce qualitative differences. If you sow two pinches of poppy seed and thin out the seedlings of one group, while leaving the others crowded, the plants in the first group will be considerably larger than those in the second. However, this improvement owing to favourable conditions is only temporary and is not transmissible through seeds. Seeds taken from the crowded starved group will be capable of producing plants as vigorous as those in the first group under similar conditions. The fact that plants under cultivation are generally larger than those in the wild may quite simply be due to the improved environment. It is true that after many years in cultivation gigas forms may appear, but these, although they may not be dissimilar in appearance, are not the same as well-fed plants. Once the gigas plant with its extra-large cells has developed, it will breed continuously with these extra large cells, even under conditions where the plant may be starved and stunted. It is not only in the size of the cells that gigas forms differ from wild forms; the chromosomes are also larger and very often the cells increase at a far greater speed. The leaf, or any other part of a gigas plant, contains not only larger cells than the wild type, but also more of them. Indeed in some gigas forms, the gigantism seems to be due entirely to the increase in cell numbers.

Although cytology and genetics have not helped the flower breeder as much as some workers have claimed, they have prevented the waste of time that would be caused by attempting impossible crosses. I remember once thinking that a good hybrid plant might be made by crossing *Sidalcea* with *Althaea*; the plants both belong to the *Malvaceae* and look similar, although *Althaea* is considerably larger. Unfortunately the haploid count of *Sidalcea* is either 10 or 13, while that of *Althaea* is 7. Moreover most of the *Althaea* spp. are hexaploid or even more polyploid and it would evidently be a waste of time to try to combine the two genera.

To try to envisage what the flower breeder has to do, let us consider a hypothetical case. There are two Rhododendrons, each with desirable features that would be very attractive if combined. These are *R. augustinii*, a hardy shrub with, in its best forms, very attractive blue flowers and *R. maddenii*, a tender species with large, very fragrant white flowers. How agreeable it would be to combine the two and produce a hybrid with blue fragrant flowers.

Although the number of Rhododendron hybrids is very large, we do not know such a great deal about how their characters are transmitted. Hybridization has not proceeded much further than the nineteenth-century method of choosing two promising-looking parents and hoping for the best. Rhododendrons have been divided into two main groups, distinguished by their leaves. Those that have the underside of the leaves covered with scales are known as Lepidote, while the series that have these scales replaced with hairs are Elepidote. There is a hybridity barrier between lepidote and elepidote Rhododendrons, that has only once been breached. Fortunately both our selected parents are lepidote, so there should be no bar to hybridizing them. Prior to 1950, very few of the species had had their chromosomes counted, but nearly 300 were published in the Rhododendron Year Book for that year, including our two species. The result has proved rather surprising. Nearly all the elepidote species are diploids with $2n=26$ and so are many of the lepidote species, but *R. augustinii* is a tetraploid with $2n=52$, while both tetraploid and hexaploid forms of *R. maddenii* are in cultivation. In order to get a fertile plant, therefore, from our cross, it will be necessary to obtain the form of *R. maddenii* which has 52 somatic chromosomes; that with 78 would give us a pentaploid plant which would almost certainly be sterile, so that if we did not obtain what we wanted from the first cross, we could advance no further. With the right parent, we can breed further generations.

We do not know about the relative dominance of colours in the lepidote series, but David Leach in his *Rhododendrons of the World* (London, Allen and Unwin, 1962) gives a probable table for the elepidote species. If the lepidotes behave in the same way, blue would be epistatic to white. The genetic basis and inheritance of fragrance is, presumably, being investigated, but no very convincing results have yet, so far as my knowledge goes, been published. In the genus Rhododendron fragrance is associated with light-coloured flowers. *R. maddenii* has been crossed with the red *R. cinnabarinum* and the resultant hybrid ('Royal Flush') has no fragrance.

Plate I
New Ericas

Plate II
Yellow Moutan (Tree Paeony)

We make our cross, but which do we use as pollen and which as seed parent? The nineteenth-century breeders would have used *maddenii* as the pollen parent, assuming that hardiness or the lack of it was a maternal quality. If we have space, we can make the cross both ways, but if space is at a discount, we had better follow the nineteenth century. The offspring must now be grown on and flowered. This will take about five years. We may get what we want among our seedlings, but if we do not, we can probably expect a pale mauve flower, which may or may not be scented. We select the best looking of these for future breeding, although it is just as likely that the less attractive plants may contain the genes we want. We can cross these amongst themselves, and we can back-cross them to our two parent species. We now have three strains going: (*augustinii* X *maddenii*) X (*a* X *m*), *augustinii* X (*a* X *m*) and *maddenii* X (*a* X *m*). We need another five years before these flower and we also need rather a lot of room, as we probably have a good many plants. We may get what we want as a result of this second generation, but if we don't we may have to back-cross once again. There seems little reason to suppose that the third generation of the original cross will produce what we want, if the second hasn't, but a further crossing with one of the parents, if we wish to enhance either colour or fragrance may be called for. It is possible therefore that we may get what we want after fifteen years or we may regretfully have to conclude that the hardy fragrant blue rhododendron is unattainable. If we were dealing not with shrubs but with annuals, the whole thing would take three years, which explains why annuals are bred more extensively than shrubs. There may be rhododendron breeders who have bred second generation crosses, but usually one breeder produces the hybrid and another breeder may self it subsequently. There are, however, examples of back-crossing to one parent and this seems to be productive of good results. As a general rule it still seems safe to say that in many fields we are as empirical as our ancestors, in spite of our increased knowledge.

CHAPTER II

# Doubleness and gigantism

In the last chapter we dealt with the broad biological principles of plant breeding. Before going on to examine in detail man's work on certain plants, let us consider two of the most sought-after phenomena which occur in cultivated plants: doubleness and gigantism.

Gigantism is one of the commonest phenomena in cultivated plants. There is a natural tendency when one sees outsize plants to assume that polyploidy, an increase in the number of chromosomes per cell, has occurred, but in fact giant forms can occur without any increase in ploidy. The phenomenon is more common in grains and vegetables, but the modern florists' form of *Cyclamen persicum* was originally an example of gigantism. When this takes place, every individual cell in the plant is two or three times as large as in its wild ancestor, with the result that the plant is considerably larger in all its parts. This is obviously gratifying in plants such as Peas or Carrots. The giant Cyclamen, with that curious synchronicity that so often characterizes changes in plants (the 'Spencer' Sweet Pea is another example) appeared in 1870 both in Germany and in Great Britain. The plant had been in cultivation on a fairly large scale for some forty years and larger forms had already been selected, but the final leap forward was spectacular. The position is now not so simple as tetraploid forms have occurred, notably in the red and salmon shades, and the modern Cyclamen is not entirely a product of gigantism pure and simple. As a matter of fact, although gigantism can be expected in cultivated plants it is not very common in plants grown for ornament. Modern breeds of such plants as Snapdragon and Hollyhock appear to be giants when compared to their wild ancestor, but this is the result of giving the plants a culture that allows them to grow to their utmost, as well as the result of generations of selecting the most promising forms. If you grow a box of modern Snapdragons and do not prick them out, the crowded and starved plants will be like wild forms. If you do the same thing with the Cyclamen, the plants will not be very floriferous, but they will have the characteristic wide petals of the giant form and will not revert to the long narrow petals of the wild *C. persicum.*

Among florists, and to a certain extent among gardeners, plants with double flowers are particularly prized. There are different kinds of doubleness, but the form in which it is most commonly met is when the reproductive organs, the stamens and/or the carpels, fail to develop and appear as petal-like. It is not very often that one finds a double plant in the wild and the reason is plain: if the carpels have become petaloid the plant will be unable to set seed, so this particular form will be unable to reproduce itself. If the carpels are unchanged, but the stamens are all petaloid, fertilization is possible, but the double flower will have no attraction for the insects, but for whom pollination is unlikely to take place. In the garden or nursery these flowers can be pollinated artificially and the double strain can be perpetuated, but this will rarely happen in the wild. Double flowers are undesirable from the point of view of Natural Selection and so tend to die out.

The most usual form of doubleness is when the stamens, either entirely or in part, are transformed to petal-like structures. In some cases this seems to be genetic; the gene that is responsible for the formation of stamens is suppressed and the gene that produces petals takes over. In the case of *Primula sinensis* there are two genes whose mutants cause doubleness. The gene that Haldane and De Winton* call *X* normally causes the formation of stamens and styles and its mutant causes the appearance of sterile double flowers; on the other hand the gene *M* suppresses an extra whorl of petals being formed and its mutant gives double flowers, that are fertile.

* *Op. cit.*

Dr A. L. Hagedoorn in his *Plant Breeding* (London, 1950) relates an interesting experiment conducted by Meunissier. The white-flowered Mexican Poppy, *Argemone platyceras,* was crossed with the yellow *A. mexicana.* The first generation resembled the parents in appearance but was intermediate in colour. These plants were interfertilized and the second generation showed surprising variability; plants with pink, salmon and pale lilac flowers appeared, some with petals of unequal length, others with laciniate petals and a few of the offspring were double. Dr Hagedoorn suggests that *A. platyceras* has a gene *A* which ensures singleness of flower and is dominant over the recessive gene *a.* In *A. mexicana,* two *aa* genes are present and it is a different gene *B* that causes the formation of stamens, which the recessive *b* will replace with petals. The first cross gives *AaBb* or a single flower all the time, but in the second generation one plant in sixteen will carry *aabb* and have double flowers. It would seem, from the frequency with which double flowers appear, that the *Papaveraceae* do have this recessive gene which causes doubleness. There is a long tradition of double poppies and they can now be expected fairly confidently from seed. It is, as usual, the stamens that are converted into petals, and the carpels are unchanged, so that the flower can be fertilized and will set seed. When double flowers are fertilized by singles, the $F^1$ generation may be mainly single, but the doubleness often appears as a simple Mendelian recessive in the next generation.

Many Paeonies show doubling. The Chinese Paeony, *P. lactiflora,* has been cultivated for more than two thousand years and is known in many forms. The numerous stamens become petaloid, sometimes resembling closely the true petals and sometimes becoming quill-like to give an anemone-centred effect. All the stamens may be affected or only the outer ones, so that a semidouble fertile flower results. Another famous double paeony is the early-flowering double red of cottage gardens. This is always assigned to *P. officinalis,* but it does not seem to agree with the description of any wild form of this plant. The leaves of the wild *P. officinalis* are divided into more and narrower segments than the old double paeony, the underside of the leaves are villous in the wild plants and quite glabrous in this old garden plant. The matter is complicated by the doubt as to what *P. officinalis* meant in Linnaeus' day. In the *Species Plantarum* (1753) Linnaeus refers to two forms of *P. officinalis,* which he calls *mascula* and *femina.* In 1768, in the eighth edition of his *Gardener's Dictionary,* Miller adopted Linnaeus's binomial system and referred to *P. mascula* and *P. femina.* He states of the latter that there are several varieties with double flowers, which are cultivated in the English gardens. He also states clearly that the leaves are hairy on the under side. Miller suspected that the 'large double purple Peony' was a form of *Paeonia peregrina.* This is a native of the Balkans and western Asia and its leaves are very similar to those of our common double paeony, but the more recently introduced forms flower considerably later. De l'Obel* in about 1570 referred to our paeony as *P. promiscua neutra.* Could the adjective *promiscua,* I wonder, indicate a hybrid origin? *P. officinalis* and *P. peregrina* are known as both diploids and tetraploids. Most *Paeonia* species hybridize fairly easily and it may well be that our old garden paeony is a hybrid with *P. peregrina* as one parent. The true officinal paeony, used by the monks in medicine, was not our modern *P. officinalis*, but *P. mascula* (*corallina*) which is found naturalized in many parts of Europe near the sites of old monasteries. Miller also refers to another double Paeony, which arose from seeds 'brought from the Levant'. This he called *P. tartarica.* The leaves were downy on the underside and the flowers were a bright red, slightly smaller than those of *P. femina.* There are some flowers of which the origin is doubtful. The Tulip is pre-eminent among these, but there is also the perpetual Carnation and the Auricula, for which parentages have been suggested but not demonstrated and it would seem that we might add our old Double Paeony to this list.

Although it has not been cultivated for quite so long as the Chinese Paeony, there are records of the Tree Paeony (*P. suffruticosa*) being cultivated in China since the Sung dynasty and during this long period a large number of double and semi-

* M. de l'Obel, *Plantarum seu stirpium historia* (1570).

double forms have been evolved.

The double forms of the Primrose, *Primula vulgaris,* are nearly sterile, but a little pollen can usually be found and used on singles to breed further doubles. There is an additional problem here for the plant breeder. Most of the genus Primula have their reproductive organs in two forms. There are the pin-eyed plants, which have a long style and short stamen and the thrum-eyed, which has a long stamen and short style. In order for seed to be produced it is necessary for pollen from a long stamen to be placed on a long style or pollen from a short stamen on a short style. This device ensures cross-pollination which is genetically desirable (although many plants do without it). With a double flower and very little pollen, it cannot be easy to know for certain whether your original flower was pin or thrum, and in the first pollination some of the pollen will be wasted. Afterwards the breeder will presumably have made a note as to which his flower was. Heterostyly, in which the styles and stamens are of different lengths in different plants, is found in a certain number of genera, but, with the exception of Primula, they have not yet come into plant breeding to any great extent.

Complete doubleness with all the reproductive organs petaloid is not very frequent. The most conspicuous example is the Stock (*Matthiola incana*). This flower is completely sterile and the double Stock owes its survival to a strain that has been known for over three hundred years. The doubleness appears to be due to a recessive gene, and in this strain, known as the eversporting, the pollen cells in the single plants suppress the gene for singleness. If we designate the dominant gene for singleness as *A* and the recessive gene causing doubleness as *a,* we can see what happens. In a normal plant with *Aa* in both pollen and ovules, the number of possible different combinations is *AA*, *aA*, *Aa*, and *aa*. Since *A* is dominant, this means that the ratio of singles to doubles is 3 : 1. With the eversporting strain the pollen is *aa* and the ovules are *Aa,* so that the resulting plants tend to be either *aa* or *Aa*; that is the proportion of doubles to singles is equal. If you go back for a minute to the normal plant, of the four possible combinations of genes, *aa* is double and sterile; *AA* has lost the recessive *a* gene altogether and will continue to breed singles, but the *Aa* and *aA*, although similar in appearance to the *AA* plant, will continue to produce one double in four.

However, as these plants are indistinguishable in appearance from the *AA* plants, it means that if all the seed of the single plants is sown the amount of doubles will be considerably less than the expected one in four, as the *AA* plants will all be breeding single. Had it not been for the recognition of the eversporting strain, it is unlikely that the double Stock would have continued as a garden flower. The first Stocks were biennial and it would have been possible to perpetuate the double flowers by vegetative propagation, but such plants tend to get lost as gardens decay, as nurserymen die or as wars turn men's attention to other matters.

When the carpels as well as the stamens become petaloid, the plant can only be propagated vegetatively. Some plants can be propagated in this way for centuries without coming to any harm. The double daffodil Van Sion has been known since the seventeenth century and the sterile Dutch Yellow crocus for even longer. Other stocks may get weakened through a virus or by fungal disease. The double Parma Violet barely remains in cultivation, although there may well be healthy stocks still in cottage gardens round Naples and many double flowers that were known in the days of Parkinson and Miller are now lost. No one now has seen the double-flowered Crown Imperials, which Miller said were the most valuable. Double Hyacinths had a long season of popularity, but became very scarce. They are being bred afresh nowadays.

Some doubleness seems to be an equivalent of gigantism, and if the plant is starved and side-blooms are taken off it is possible to get plants that resemble their single counterparts. In some cultivars of the Chinese Paeony, the centre flower of the cluster will be fully double and have no stamens, while the lateral flowers will be single or semi-double. It may well be that rich living encourages some plants to produce extra petals and that this process can be reversed by starving the plant; a treatment that is often resorted to with Carnations. The double Carnation usually holds a fertile stigma and no stamens. One way of breeding new Carnations is to use as the seed parent the very double flowers that burst their calices, with a single flower as the pollen parent. The result is generally a large proportion of flowers that are sufficiently double to be attractive, but not so large that the calices burst. However if you have a particularly good double plant from which you wish to breed, you can reduce the doubleness by starving the plant. Since, however, the doubleness will be reproduced in the progeny, it must have some genetic basis as well. It seems an odd form of a genetic character that can be reversed by hard treatment and one is led to the conclusion that the causes of doubleness are still largely unknown.

A quite separate form of doubleness occurs in the *Compositae*, among which are listed such well-known flowers as Dahlias, Chrysanthemums and Zinnias. This, one of the largest of plant families, has evolved a sort of co-operative amongst its flowers. There are two sorts of flowers, well exemplified in the common Daisy of our lawns; the white ray florets surround the mass of small yellow disk florets. Some composites such as Groundsel have only disk florets; some, like the Dandelion, have only ray florets and so appear double; but most have a combination of the two. One can find something similar in some of the *Umbelliferae*, in most *Hydrangea* and some *Viburnum* spp. Here again you have an outer ring of conspicuous florets, which are generally sterile, while in the centre are very many small fertile flowers. In this way the plant gives the impression of one very large flower, which will attract pollinators, but it conserves its energy by producing a relatively small number of petals. As a result of mutation, some of these plants have only the outer sterile flowers preserved and you get such plants as the Snowball Tree (*Viburnum opulus sterile*) and the garden Hydrangea (in which the large florets are not sterile but bear pollen). Among the *Compositae* the outer ray florets are sometimes sterile, sometimes hermaphrodite and sometimes unisexual, but it is usually quite easy to breed plants that are entirely composed of ray florets and thereby give the appearance of a double flower, although plants with disk florets may be necessary to obtain seed. Here again it has been found that starving the plant will inhibit the formation of supernumerary ray florets and give 'single' flowers. In most composites the ray florets are unisexual and so can be used in pollination and in some cases they are hermaphrodite and will produce flowers of their own accord.

What appears to be a form of gigantism is observed among some composites. This is the so-called Anemone-centred flower. Here the ray florets are more or less unchanged, but the disk florets are considerably enlarged, so that each floret is distinctly separated from its sisters, and these florets may either take the colour of the ray florets or take up some other colour. Some of the Anemone-centred Chrysanthemums have ray florets of one colour and the centre in another. Generally the disk florets are yellow to give an impression of stamens to the pollinators, although other colours are known. The disk florets of some *Rudbeckia* spp. are a dark olive or nearly black. The term Anemone-centred has survived the form of Anemone from which it was named. From the late sixteenth century to early in this century forms of *Anemone coronaria*, the Poppy Anemone, were grown in which the stamens had been transformed into quill-like processes of the same colour as the tepals, but considerably shorter. In modern double anemones, such as the popular 'St Brigid' strain, these petaloid staminodes are the same length as the tepals and the plants appear fully double. The staminodes have become longer and broader.

The case of the Begonia is rather curious. The plant is monoecious, having male and female flowers on the same plant, and the double flower is invariably the male. This means, of course, that it is the stamens that have become petal-like. It seems relatively easy to obtain this phenomenon in the tuberous species and it has been obtained in some of those with fibrous roots. I know of no record of doubling among the rhizomatous Begonias. Many of these are grown for their foliage rather than their flowers, but a plant such as *B. manicata*, or the various *coccinea* hybrids, which are grown for their flowers, show no tendency to double. Like some other cases we have been discussing, it is possible to reverse the tendency to double in the male flowers of the tuberous begonia, by starving the plant.

Some families appear to be more prone to doubling than others. Among the Monocotyledons doubling is not a particularly common phenomenon. It occurs only sparingly in the *Liliaceae*. In the seventeenth and eighteenth centuries there were double forms of the Madonna Lily (*L. candidum*), the Orange Lily, (*L. croceum*) and both coloured and white double Martagons. These have now all been lost to cultivation. Miller in his *Gardener's Dictionary* says that the double Madonna Lily was liable to rot before it opened and would only give a satisfactory display under glass. The only double lily in cultivation nowadays is the double *L. tigrinum*. Now that Lilies are being bred on a very large scale, it is possible that more doubles will turn up, but since they are not very attractive they may well remain unpropagated. The double Hemerocallis is a rather graceless plant and has not proved very popular. Double Colchicums are attractive, but occur very rarely and have always been scarce plants. The *Amaryllidaceae* show more double flowers. There are numerous double Narcissus, some dating from the seventeenth century. The double Snowdrop produces some pollen and appears to be dominant over the single wild type. As it does not use up energy in producing seed, it also makes many more offsets than the single plant and increases with great rapidity. About ten years ago double Freesias were produced. These are the first members of the *Iridaceae* to show doubling; the so-called double forms of *Iris kaempferi* have petaloid styles and, occasionally, an additional set of petals, but they continue to bear stamens and styles. The double Freesia was a most spectacular flower, very double indeed and looking quite different from the graceful single flower. It gave the impression of being so double that it would probably be sterile but if the consequent difficulties can be overcome it is clear that it is potentially a first-rate plant for the cut flower trade. Most double flowers appear rather clumsy compared to the singles, but the double Freesia did not look like a Freesia at all, but more like a Gardenia, so that no unpleasant comparisons were aroused. Otherwise the *Iridaceae* have not shown doubling, although one would have expected this to have turned up among plants as extensively cultivated as the Gladiolus. If, however, it has occurred, it has neither been reported nor propagated. Although occasionally Orchids may be seen with additional petals or lips, nothing like doubling has ever occurred in the *Orchidaceae*.

The picture is different when we turn to the Dicotyledons. Here there are few families in which doubling has *not* been observed, though it may not always have been encouraged. In the case of the *Papilionaceae*, for instance, the flower is literally doubled, bearing two standards, two keels and, occasionally two pods. But the result is not attractive and the double form has not been cultivated for that reason.

It would seem then that doubling is a rather mysterious phenomenon. In some circumstances it appears to be due, without any question, to a mutation and in a few cases it seems connected with polyploidy, but in others it appears as a phenomenon that is reversible. Once a mutation has occurred, one would expect it to be permanent. If one only got a flower with fewer petals as a result of giving the plant harsh treatment, the theory that all doubling is due to a mutation would hold, but this does not always happen. Sometimes the plant suppresses the extra petals and produces fertile stamens instead. In spite of this the offspring of such plants will often show doubleness. It looks rather as if we have a state that is transmissible genetically, but which is also dependent on external stimuli. This is not usually observed in other phenomena. Where, as in the case of Stock, there is no doubt of the genetic character of the doubling, the doubling cannot be nullified however much the plant is starved. It looks as though some doubling may be rather similar to a disease like gout; this is reported to be exacerbated by over-indulgence in eating and, more particularly, drinking and the susceptibility to the complaint is said to be inherited through the male line. On the other hand it can be, if not cured, alleviated. Possibly doubling is an inherited tendency that can sometimes be alleviated. As we have seen, not all doubling is the same and the explanation for one form may not cover another.

There does appear to be a different attitude nowadays towards double flowers. They are more spectacular and longer lasting than the singles, but they are now considered to lack the grace of more natural blooms. However, the double Gardenia has been in cultivation so long that very few have ever seen the single type; and the double Tuberose is still to be seen commonly, while the more attractive single plant is quite a rarity. To most people the word Rose immediately calls up a double flower. No one today wishes to grow single tuberous Begonias. Chrysanthemums, Dahlias, Zinnias and many other composites are more popular when the capitolae are composed entirely of ray florets and the single Carnation is despised. We certainly cannot say that the double flower is unpopular, although its attraction may be less than it was.

CHAPTER III

# The development of plants in China

*Paeony – Azalea – Camellia –* Primula sinensis *– Chrysanthemum – China Aster – Chinese Pink*

When European gardening came into existence, growers did not have to rely only on wild plants, whether native or collected in other parts of the world. Many flowers introduced into Europe, especially during the eighteenth and early nineteenth centuries, had been in cultivation in the Far and Middle East for hundreds of years and had been improved by constant, if unscientific breeding. I shall therefore begin my discussion of cultivated flowers with a brief account of some plants that originated in China and Turkey and which, on their first appearance in Europe, were already man-made flowers.

There is one important omission from these chapters: the Rose. China's chief contribution to our present-day garden Rose was the remontant form of *Rosa chinensis,* but the whole history of this genus is so complex that I have thought it best to defer discussion of all the ancient species to a separate chapter (Chapter V).

China has by far the longest known history of gardening and so it is not surprising that very many of our most popular plants are of Chinese origin. Moreover many plants that we associate with Japan came to that country from China. The most recent study of this subject is *The Garden Flowers of China* by H.L.Li,* to which this chapter is heavily indebted.

* The Ronald Press, New York, 1959.

It would appear that Chinese gardening was at its most developed in the eleventh century during the Sung dynasty. Evidently there have been developments since that date, but they appear to have been further refinements rather than new breaks and the main guide-lines may well have been laid down nine hundred years ago. The plants that have been intensively imported and cultivated in Europe include the herbaceous and tree Paeonies, the Chrysanthemum, the Camellia and *Azalea indica*; and, of course, our races of remontant roses all spring from forms of *Rosa chinensis*. *Prunus mume*, the Japanese Apricot, has long been popular in China and many forms are known there, but it is too tender for many parts of Europe and is comparatively little known in the West. The foregoing together with *Primula sinensis* are the species that have been most developed in cultivation, but there are, of course, numerous other plants originally cultivated by the Chinese which now adorn our gardens. Indeed some of these would now be extinct, had they not been cultivated. No one has found a wild Ginkgo nor a wild *Malus spectabilis* nor a wild *Primula sinensis*. European collectors such as Forrest, Wilson and Farrer and many others, have enriched our gardens with flowers that the Chinese themselves had never used, but, with the exception of *Hemerocallis*, the types of flower with which this book is concerned had already been developed by the Chinese for many years before they were introduced to Europe.

The recorded history of Chinese gardening dates back far longer than any comparable account in any other country. If we read that some plant has been recently introduced in China, we are liable to find that it has been in cultivation for five hundred years. The earliest reference to the Paeony (*P. lactiflora*) is in the Book of Odes of the 5th century B.C. In the Han dynasty the emperor Wu-ti (141-85 B.C.) created a large garden in his capital of Changan, while in the late sixth and early seventh centuries, the Sui emperor Yang-ti built the most extensive imperial gardens in the whole history of the empire. When we consider these hundreds of years of systematic selection and cultivation it is not surprising that the development of garden forms should be so advanced, nor that the Chinese should still be so pre-eminent in horticulture.

Although such fruit trees as the peach had spread westwards in early times, it was not until the eighteenth century that Chinese flowers became known to Europe, although the existence of many were familiar before by their representations on paintings and porcelain.

The earliest introduction was *Camellia japonica* which was imported in 1739. Although most early authors give China as its origin, it may have been imported from Japan, otherwise why should Linnaeus have called it japonica? *Paeonia lactiflora* was received in 1784, ostensibly from Siberia. Loudon gives the same year for the introduction of the double varieties known as *whitleyi* and *humei*, while Sweet gives 1808 and 1810 for the years of their introduction. The two authorities differ also over the introduction of the Chrysanthemum, Loudon giving the improbable date of 1764, while Sweet says 1790. According to the R.H.S. *Dictionary*, the first importation was to Holland in 1688 but the plants did not survive. In 1789 Captain Blancard of Marseille brought from China several plants which were successfully established in France and plants derived from these were received in England in 1795. One of the parents of the modern Chrysanthemum, *C. indicum*, arrived considerably later, probably in 1820. It seems generally agreed that the first Tree Paeonies were received in England in 1789. Importation was difficult for many reasons and the following account in the *Floricultural Cabinet* of 1842 by a writer calling himself Clericus is interesting.

'The introduction from China of Moutans of any description is attended with difficulty, for of the plants which are put on shipboard in China to be brought to England very few live to reach their destination. With the exception of the Azaleas they seem to bear a long voyage worse than any other productions of the Chinese gardens which we have hitherto obtained.

'Large quantities of flowering plants, closely laid together in open packages, without mould to their roots are annually brought in the course of the winter from distant parts of the

Chinese empire to Canton. These, notwithstanding this exposure, blossom in the ensuing spring; but either from the climate not agreeing with them, or the treatment they receive being unsuitable, the state of those which survive to the autumn is such that they are not fit for removal with any chance of success. After their first blossoming at Canton these plants never flower again, but dwindle and decay; and from this cause the captains of the British Indiamen, which leave Canton in the winter season, are unable to obtain any which have been proved to be of the more desirable kinds. Their purchases are necessarily made from the stock brought into the market in the manner above mentioned, in which the varieties most wanted are either very rare, or only sold to the Chinese, and are, besides, not very easily distinguishable whilst divested of their foliage; so that the living plants which do arrive in England usually turn out to be the sort which we have had here longest as well as in most abundance, and which it may be presumed is the most common in China, or at least at Canton.'

When we think that these plants, divested of all soil, had to survive a journey on the sea for many months, we may well be surprised that any were received alive.

## ❀ *Paeony*

The first mention of the Moutan or Tree Paeony (*P. suffruticosa*) is by the poet Hsieh Kang-Lo who lived during the Tsin dynasty (A.D. 265-420). The popularity of the flower reached its apogee in the eleventh century and there still survives a monograph on the species written by Ouyang Hsiu, who lived from 1007 to 1072. In this monograph *A treatise on the Moutan Paeony of Loyang* it is mentioned that more than ninety varieties were known to be in cultivation, although only twenty-four are described in the monograph. He tells us that stocks were propagated by budding on to wild *P. suffruticosa* and also by grafting on to *Paeonia lactiflora,* as is done to this day. He relates that, in order that the flowers might be viewed from an upstairs window, a graft was successfully made on the Chinese mahogany tree (*Cedrela sinensis*). This sounds improbable, but since there are apparently many instances of roses and Citrus plants being grafted on to pomegranates one would not like to say that it was impossible. However we have not heard of it being done more recently. It does not seem to have been very long before double flowers made their appearance. Although what we regard as the typical wild form is white, with a purple blotch in the centre, the most popular of the early varieties were red. Now we approach a mystery. After the reds had become common, the most expensive were described as yellow. This seems a good place to introduce an extract from Ouyang Hsiu's work:

'In Loyang every family has the flower, though few of the plants become tall because without grafting the plants do not grow well. In the spring some people bring seedlings from the hills to sell to the citizens. These are planted in rich ground and are grafted in the autumn. The most famous grafter is named Men the gardener. A single graft of Yao's Yellow costs 5,000 pieces of cash. A contract is signed in the autumn to be given to Men and the sum is paid in the spring when the flowers are produced and proved true to name. The people of Loyang are particularly fond of this variety and are unwilling for others to propagate it. . . . When Wei's Flower made its first appearance, a bud cost 5,000 pieces of cash and it is still valued at 1,000.'

Now what were these yellow Tree Paeonies? The Chinese also had varieties of *P. lactiflora* that they called yellow (see below, p. 60). Since the introduction here (to Europe) of *P. lutea* in 1900, crosses have been made with *P. suffruticosa* to give yellow Tree Paeonies, but there seems no evidence to suggest that the Chinese cultivated *P. lutea* at all, although it is, of course, a wild plant of western China. Mr Li in his book says that new forms of *P. suffruticosa* were developed from seed, although no records show that deliberate crossing was resorted to. He does record, however, that it was believed that chemicals applied around the roots would affect the flower colour. The formula appears to be lost whereby the white Moutan was transformed into Ou's Green, which was even more esteemed than Yao's

Yellow. But we are told that extract of saffron would give the plants a reddish colour, extract from the roots of *Lithospermum officinale* would make the white Moutan purple, while by drying and powdering the roots of *Atractylis ovata* (a kind of thistle) every petal would have a yellow stripe down the centre. It all sounds wildly improbable, but I have no reason to assume that anyone has put it to the test and I cannot think of any other way in which a yellow Tree Paeony can have been produced. Of course it is possible that the word which is translated 'yellow' is really only a deep cream, but failing this, we are left to assume that crosses were made with *P. lutea*, or that some form has been lost; and in any case the appearance and real colour of Yao's Yellow seems to be unknown. It may be depicted in some painting on silk or porcelain, but I have never seen it.

In a collection of Chinese paintings in the Lindley Library, thought to have been executed in about 1826 for James Reeves, the tea buyer for the East India Company at Canton, there are depicted both a yellow and a copper-bronzy coloured Tree Paeony. This does suggest hybridization with *P. lutea* on the one hand and *P. delavayi* on the other. The yellow Moutan in the Reeves painting is shown with a red-coloured bud, which agrees well with modern hybrids of *P. lutea* and *P. suffruticosa*. On the other hand, Robert Fortune in his *Three Years' Wandering in China* (1847) says that the yellow Tree Paeony 'was only white with a slight tinge of yellow near the centre of the petals'. Reginald Farrer repeats this statement in *On the Eaves of the World* (1917) although it is not clear whether he is speaking from personal observation.

The first Moutan to reach Europe was received at the Royal Gardens at Kew (later the Royal Botanic Garden) in 1789, but did not flower until 1793. At that time all plants that we now regard as cultivars were treated as varieties and given varietal names. The first importation, which was at the instigation of Sir Joseph Banks, was given the epithet *banksii*. The original flower was fully double and about nine inches in diameter and the colour is described as blush with a white edge and purple staining at the base of the petals. It sounds rather like the form that is most commonly met with all over Europe, but it was noted that the leaves had red petioles, which is not the case nowadays. Presumably seed was gathered as single and semi-double forms were also known.

In 1794 and 1795 two more forms were imported (*rosea semiplena* and *rosea plena*) which differed from each other, in spite of the similarity of their names. *P.s. rosea semiplena* was comparatively small-flowered and had deep pink petals and a rose-like fragrance, *rosea plena* was larger, not so fragrant and the interior petals were long, narrow and jagged. The cultivar '*Anneslei*' was raised from seed, though whether imported or collected from existing plants is not stated. It was small-flowered, the blooms were only 4½ inches across, 'almost single' and of a rich purplish-pink with a darker base which prolonged itself in a stripe up the centre of each petal.

In 1802 Captain James Prendergast in the *Hope* East Indiaman brought some plants for Sir Abraham Hume, who, together with his wife, was an enthusiastic gardener. One of these plants, which bloomed for the first time in 1806, was given the name *papaveracea,* but it is now known that this was the original wild form. Possibly the imported plant had been grafted on to a wild *P. suffruticosa* and the graft failed to survive. The flowers were single, sometimes as much as a foot across, with white, slightly fimbriated petals with a deep purple star in the centre. 'Clericus', writing of this plant in 1842, says 'the flowers in some seasons, and especially of late years, have become semi-double'. I imagine that he is referring to selfed seedlings raised from this plant, as he says that there were now many of them. Among other importations of Sir Abraham was a cultivar named 'humei' after him, which was red and double. There were other purplish double flowers and one that was nearly white. When Robert Fortune went to China in 1842 he brought back as many as forty varieties and these were subsequently developed in England and France to give the varieties to be found in catalogues today. In recent years the aeroplane has allowed further varieties raised in Japan to be imported. The Japanese cultivars tend to be semi-double, whereas the European breeders had previously concentrated on producing fully double flowers. However this tendency had already begun to wane before the Japanese importations and many of the more recently introduced European cultivars have been semi-double. It would seem that there has been no significant development in the last eight hundred years. It is interesting to note that, following the suggestion of a Mr George Anderson, the great botanist D. Don agreed that the so-called variety *papaveracea* was 'the normal form of the species'. This must have been in about 1837. It appears from Loudon's *Hortus Britannicus* that a large number of forms were raised from seeds of *papaveracea*, presumably crossed with other cultivars as the colours included reds and pinks. Most of these flowered first in 1832.

The well-known Chinese Paeony, *P. lactiflora*, was, as we have seen, cultivated in China for far longer than the Tree Paeony. It appears to have been the subject of various monographs in the tenth and twelfth centuries: the earliest one enumerating thirty-three varieties has unfortunately been lost, although the author's name is preserved. The first complete one, written in the latter half of the twelfth century, describes thirty-nine varieties and was written by Wang Kuan. The wild plant is a native of Siberia, Mongolia and northern China. Apart from its beauty, an extract from the roots was used medicinally. There were both single and double forms and there was a yearly trade of sending plants from the north to the south, where they flowered for a single season in the same way as the Moutans.

Here again we find rather perplexing references to yellow forms. It is possible perhaps that the Chinese had bred the so-called Japanese or Imperial forms, in which the stamens are converted to yellow or purple petaloid staminodes. If this was done with a variety of white petals, I suppose the resultant flower could be described as yellow. I have not seen these petaloid staminodes in Tree Paeonies, otherwise we could postulate a similar explanation for the yellow forms of these. If this is not the explanation we are still faced with a mystery. There are four yellow Paeonies known. Two of these, *P. wittmanniana* and *P. mlokosewitchii*, are natives of the Caucasus, *P. wittmanniana* also extending into Persia. The other two *P. lutea* and *P. potanini* var. *trollioides*, are Chinese natives, but being members of the suffruticose section *Moutan*, would probably not hybridize with the herbaceous species. Cytologically there is no reason why they should not, as both have the same chromosome count ($2n=10$). I would imagine that if any cross were possible it would have been reproduced in our own time. Moreover we have no knowledge that deliberate crosses were made by the Chinese at that early date, although their knowledge seems to have been such that it is by no means impossible.

*P. lactiflora* was first received in England in 1784 (the date of 1584 in the first edition of the R.H.S. *Dictionary* was a misprint) and this seems to have been the wild single white form. It was received not from China, but from Siberia. Other forms followed, including a blush-coloured one, but the first real impetus to its popularity was when a nurseryman of Fulham, R. Whitley, imported a double white form from China, named after him as var. *whitleyi*. Two years later Sir Abraham Hume at Wormleybury imported a double crimson (which, like the cultivar of *P. suffruticosa* already referred to, was given the name *humei* after Sir Abraham); while Reeves imported a double pink in 1822. These seem to have been sufficient to give the basis for all future breeding. I am not sure when the Anemone-centred forms were first introduced, but they are the only break in recent times and they may well have been developed in the tenth century and given rise to the so-called yellow flowers.

In recent times nurserymen have attempted to cross both *P. mlokosewitchii* and *P. wittmanniana* with *P. lactiflora*. There appears to be a hybridity bar between *P. mlokosewitchii* and *P. lactiflora*, but it has been proved possible, notably by Lemoine, to cross *P. wittmanniana* with *P. lactiflora*. Unfortunately *P. wittmanniana* is a tetraploid, with the result that the hybrid is a sterile triploid. If the seedlings could be treated with colchicine and made hexaploid, the resultant plants could be crossed back to *P. wittmanniana* and a yellow *lactiflora* hybrid might be produced. But *P. wittmanniana* is not nearly so good a yellow as *P. mlokosewitchii*: it can equally well be described as a deep cream in colour and it has probably been thought that the long process would not be worth the result. It is safe to reckon on at least five years between seed sowing and the appearance of the first flower, so that if the original cross were back-crossed by *P. wittmanniana*, assuming that the polyploidy could be accomplished, a period of at least ten years would be necessary. *P. mlokosewitchii* will hybridize with *P. daurica* and with *P. tenuifolia*, but presumably the resultant hybrids have not proved commercially attractive. In any case we have heard nothing of them.

Mention should be made of the agreeable hybrids between *P. lactiflora* and various European spp. by A. P. Saunders and Sir Frederick Stern. These have made very attractive plants, but they have not, as yet, entered into general gardening.

### ❀ *Azalea*

*Rhododendron simsii*, known for so long as *Azalea indica*, would not appear to have been cultivated for so long a period in China. Mr Li, in his fascinating book, merely mentions that it is a native of China, growing on the banks of the Yangtze and Pearl rivers. This is rather far south for the enthusiastic gardeners of the Northern Sung dynasty, so it is possible that its earliest cultivation was not much noticed in contemporary literature. But we can be certain that it was known in many forms by the end of the eighteenth century, when importations into Europe began. As we have seen importation was difficult and many plants must have perished on the journey. The first plants were received in 1808 and were, apparently, the usual wild scarlet-flowered plant. No more were received until 1819 when a white and a double purple were received. According to Loudon an orange-flowered form was imported in 1824, but this must surely have been an orange-scarlet and Sweet's record of a double yellow in 1826 must be incorrect. The variety now known as *vittata*, but described as *variegata* in the early nineteenth century was received in 1824 and it was from this variety that the taking of seedlings and subsequent development began. However a cultivar with lilac flowers called 'Smithii' had already been raised by 1826, and various other hybrids were known by the 1830s. These hybrids were, of course, the result of crossing different cultivars and, although described as hybrids by contemporary writers were really more cultivars. No other species was combined with *R. simsii.* In course of time the cultivation of this popular plant moved to Belgium, where it remains an important crop to this day. Although seed is a valuable method of raising new forms, some of the most remarkable cultivars have been the result of bud sports. Of these one of the most important has been given the name *'Vervaeneana'*. It has waved petals, which are deep pink with a white edge and was introduced to commerce in 1886 by M. Vervaene, a Belgian nurseryman. 'Vervaeneana' appeared as a bud sport on a cultivar called 'Pharailde Mathilde'. This was a double white flower with the petals ornamented with red dots and stripes. It had been produced by crossing 'Königin der Weisse' a late-flowering single white flower with 'Comte Charles de Kerckhove' which sounds not unlike 'Vervaeneana', a bright rose flower with white edges and carmine stripes. It is interesting that the progeny of these two single-flowered plants produced a double-flowered offspring. We know nothing of the grandparents, but there was probably a double-flowered cultivar among them. 'Vervaeneana' is one of the few Indian Azaleas that have been crossed with hardy rhododendron species. 'Charmian' is *R. callimorphum* X 'Vervaeneana' and 'Brocade' is 'Vervaeneana' X *williamsianum.* In neither case is the influence visible and the $F^2$ generation does not seem to have been grown on.

Unlike the other plants we have been discussing, the Indian Azalea has been considerably developed in recent years. This has been done by breeding in other species. Doubtless much work has been done in China, where very handsome cultivars are to be found, but these seem to be unknown in Europe and we are better acquainted with Japanese hybrids and also with a number raised in the U.S.A. The most important of the other species used is *R. indicum*, in spite of its name a native of Japan. This is very similar in appearance to *R. simsii* and is best distinguished by the fact that in *R. indicum* the flowers are produced singly, or at most, in pairs, whilst in *R. simsii* the flowers are in clusters of from two to six. It would appear to be as variable as *R. simsii* and the Satsuki Society of Japan have named 162 forms. However some of these are certainly hybrids with *R. simsii*. The favourite forms seem to be very variegated in colour. For example 'Kokkan' has large flowers with waved petals which are white with mauve-coloured bands and darker purple spotting, while 'Misaka' has vermilion stripes on a white ground. The well-known nurseryman K. Wada has introduced hybrids that include, besides the species already mentioned, the popular Kurume Azaleas (forms of *R. kiusianum*) and *R. scabrum*. This latter is related to *R. simsii* with brilliant scarlet flowers in trusses of up to six flowers. Two other species that have been used for hybridizing are *R. mucronatum* and *R. pulchrum*. *R. mucronatum* is probably a Chinese plant, but it has never been found in the wild, although cultivated in China and Japan for some time. It has flowers in clusters of from one to three with white petals and a pleasant fragrance. It is perhaps better known under the name *R. ledifolium*. Unlike the other species we have been discussing, it is not very evergreen. *R. pulchrum* is also unknown as a wild plant and may be a hybrid between *R. mucronatum* and *R. scabrum*. Most of the forms known are double. The most usual colour seems to be rosy-purple with dark purple spots, but there are forms with bright red flowers and some that are violet-purple. It is not a very graceful plant and it seems surprising that either the Chinese or the Japanese should have thought it worth cultivating. It lacks the grace that we associate with Oriental flowers, although like the Kurume Azaleas, it is very prolific in its display. Two other species that have been used rather less are *R. kaempferi* and *R. obtusum*, var. *amoenum*. *R. kaempferi* is a variable Japanese species growing up to 10 feet high with

1. Inheritance of doubleness in flowers of *Dahlia variabilis*

2. (*Above*) Double Camellia seedlings

3. (*Right*) *Primula sinensis* c. 1826

4. Chrysanthemum c. 1826

5. Chrysanthemum c. 1826

6. Seedling Ranunculus

7. *Tulipa marjolettii*

8. Slater's Crimson China Rose

9. Parson's Pink China Rose

10. Hume's Blush tea-scented China Rose

11. Park's Yellow tea-scented China Rose

12. *Rosa*

13.
*Dianthus caryophyllus*

14.
*Dianthus plumarius*

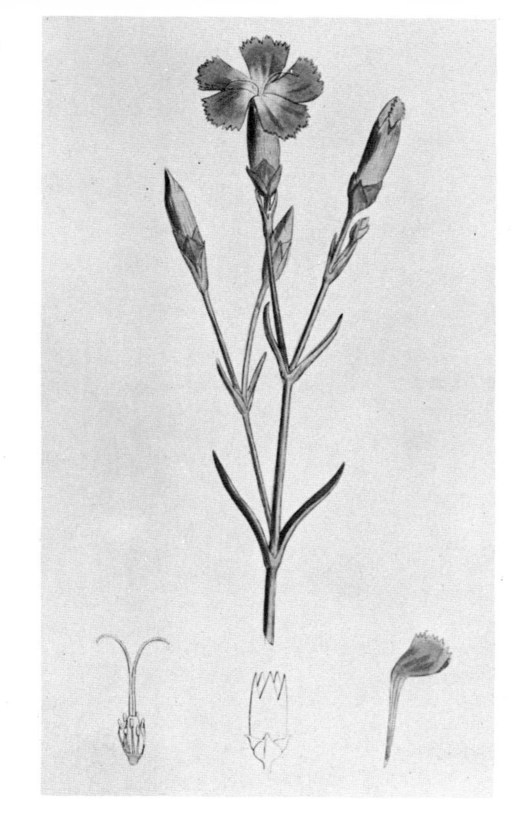

15. *Dianthus chinensis*

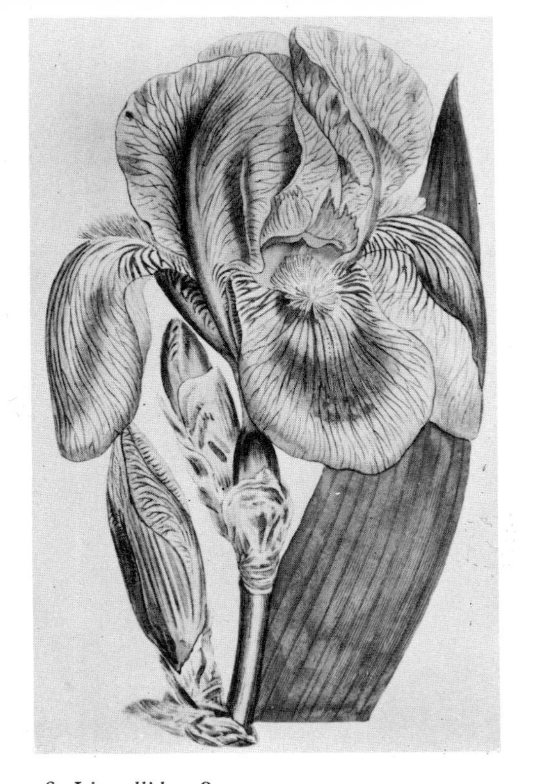

16. *Iris pallida*, 1803

17. *Iris variegata*, 1787

18. *Iris trojana*

flowers in various shades of red and also white, and with some semi-double or hose-in-hose cultivars known. It is only semi-evergreen and is often used as a stock on which Indian Azaleas are grafted. *R. obtusum* var. *amoenum* has magenta flowers, often of a hose-in-hose form (i.e. a double row of petals). This has been the basis of the so-called Arnold hybrids. The most involved hybrids are the so-called Rutherford hybrids, raised by Messrs Bottink and Atkins, East Rutherford, New Jersey. These hybrids are based on species in the *Obtusum* sub-series of *Azalea* (*RR. mucronatum* or *scabrum*), India Azaleas and large flowered Hardy Hybrids (Pink Pearl, Charles Dickens and Mrs C. S. Sargent). As far as I am aware these hybrids are not known in Europe, neither are the more numerous Glenn Dale hybrids, which include all the best-known species of the *obtusum* sub-series in various combinations. The intensive growing and hybridization has produced some colours, such as lilac and mauve, which were unknown to the European growers in the last century and there have been differences in the size of the flowers, and the arrangement of the truss. *R. obtrusum* var. *amoenum* has brought in very shining leaves that are lighter in colour than in the *simsii* or *indicum* cultivars. Presumably the Rutherford hybrids have larger flowers. As this is one of the few occasions when Europe has improved on the original Chinese work, it has seemed worth going into the matter in some detail. Of course much of the European work has been duplicated in China and Japan and we cannot show anything better than their modern cultivars.

## ❀ *Camellia*

Although the literature of tea goes back at least two thousand years, there is no written description of Camellias other than the tea tree until the late fifteenth century, which must be regarded as modern by Chinese standards. However once there is written evidence, the fact that over a hundred varieties of ornamental Camellias were in cultivation shows that their culture must have already been undertaken for some time.

There were then a number of single and double flowered cultivars of *C. japonica,* as well as of *C. sasanqua* and presumably *C. reticulata.* The most esteemed cultivar was known as 'Precious Pearl' and sounds very similar to the cultivar known as 'Mathotiana', with dark red, closely imbricated petals and a very double flower. *Camellia sasanqua* is a native of Japan and, in this case, the usual process seems to have been reversed and the plant was first developed in Japan before it was imported into China. On the other hand, the Chinese *C. oleifera* is very similar to *C. sasanqua* and may also have been developed. *C. japonica* and *C. reticulata* are both Chinese.

The Camellia was first described by James Petiver in 1702 under the name of *Thea sinensis,* but the first plants, thought to be the wild single red-flowered type of *C. japonica,* were brought to Europe by the Jesuit G. J. Kamel, in whose honour Linnaeus named the genus in 1739. Two of these plants were purchased by Lord Petre, put into the plant stove and soon perished. However his gardener Gordon would appear to have rooted a cutting or abstracted a plant for himself, which he grew in his nursery in the Mile End Road in a cool house. Here the plant flourished and set seed. In 1745 and in 1769 plants were again imported at Linnaeus's insistence. One is depicted as a pink flower of eight petals; Thunberg had seen double white and purple cultivars in Japan, but they did not reach Europe until 1802. The first doubles were a white and a striped flower. Two years later the double red was introduced and in 1806 we again find the Humes commemorated in 'Lady Hume's Blush'. At the same time a so-called 'Buff' was imported which Loudon says was white, but Sweet describes as 'pale straw'; the same colour is applied to 'luteo-alba' which Loudon agrees was pale yellow. The Anemone-centred form was also introduced in about 1806. Although importations continued for some years, enough forms were already in England (and indeed in other parts of Europe as well) for growers to start raising their own cultivars. In 1831 the firm of Chandler and Low offered '6 imported cultivars and 19 of their own raising'. 'Chandleri elegans', still a popular variety, was actually raised in 1822. Sweet's *Hortus*

*Britannicus* in the third edition, published in 1839, lists no less than 65 cultivars. Loudon, in his book with, confusingly, the same title, gives 64 up to 1839, but mentions a further 19 as not yet described. By 1830 breeding was also going on in Belgium, France and Italy and the number of named cultivars became excessive. There were a very large number that were so similar as to make their separation a matter of pride for the nurseryman rather than choice for the purchaser. Many of these cultivars have now been lost, but fresh ones are still being bred. It would seem that there has probably been little improvement on the original Chinese cultivars or on the first European raised ones. Some flowers may be larger, but in 1835 the nurseryman Ring of Frankfurt had raised cultivars which he claimed had flowers as large as *reticulata.*

*Camellia reticulata* itself was brought from China by Captain Rawes in 1822 and first flowered in 1824. For a long time this was thought to be the type of *C. reticulata,* but comparatively recently the true wild form was discovered and the original imported plant has been given the cultivar name 'Captain Rawes'. Fortune sent a double-flowered form back which flowered for the first time in 1857, but this seems to have been lost. In 1947 a number of different cultivars were discovered in an ancient garden in Yunnan and have subsequently been introduced to commerce. However it would seem that *C. reticulata* is much less variable than either *C. japonica* or *C. sasanqua* and, apart from a rather dubious record of a white form found growing wild in Hong Kong in 1859, no albino forms appear to be in cultivation. Many attempts have been made to hybridize *C. japonica* and *C. reticulata,* but until recently with no success. Nowadays considerable hybridization has been done in the U.S.A. and New Zealand. Apparently we know very little about the chromosome count of Camellia species.

*Camellia sasanqua* is a Japanese plant and, though the original importations were from China, it is probable that many of the existing cultivars are really of Japanese origin. The single white form was figured by Thunberg in his *Flora Japonica* of 1784 and was brought to England by Captain Wellbank in 1811. It was first flowered by Lady Banks. A semi-double white form dates from about the same time. In 1818 a double pink form was introduced and a double white one in either 1823 or 1824. 'Fortune's Yellow' was sent back in 1852. There are not many more cultivars known in Europe nowadays, but there are apparently very many in China and Japan. In China the plant is known to have been in cultivation for three hundred years and it was presumably cultivated for a longer period in Japan. The flowers are much smaller than those of *C. japonica* and, in England at any rate, it does not seem so floriferous. It does not seem to hybridize with either *C. japonica* or *C. reticulata.*

Here again we face a mysterious yellow cultivar. When Robert Fortune was sent to China, the yellow Camellia was first on the list of *desiderata.* His instructions read: 'It is needless to particularise at much length the plants for which you must enquire. It is, however, desirable to draw your attention to . . . Camellias with yellow flowers, if such exist.' The nearest that he came to obtaining it was the *sasanqua* cultivar already referred to, that became known as 'Fortune's Yellow', In this the outer petals were white and the inner petals were lemon yellow. From time to time rumours of the existence of a yellow Camellia in remote parts of southern Europe start up, but the plant is yet to be found. The Chinese themselves have never been very eager to sell their best cultivars to foreigners, but the Japanese have usually been pleased to do so, and we may feel fairly confident that if they had had a yellow Camellia, it would have been distributed. It would also seem probable that they would have obtained some from China if they had been easily available. One should be chary of assuming that cultivars have been lost, in view of the number of previously unknown cultivars of *C. reticulata* recently discovered in Yunnan, and it is not beyond all possibility that some old Yunnanese garden may possess a yellow Camellia.

The only occasion where effective hybridization of Camellia spp. has been carried out in Europe is the hybrid known as

*C.* X *williamsii,* raised by J. C. Williams at Caerhays in Cornwall. In this race *C. japonica* is crossed with another Chinese species *C. saluenensis. C. saluenensis* is not so very different in appearance from *C. sasanqua,* but has slightly larger flowers in white, pink and carmine. The *williamsii* hybrids have all been pink up to date and are single or semi-double. They are very floriferous but the habit is sometimes rather straggling and weak. However they have not been in existence long enough for all the possibilities to be developed. *C. saluenensis* was only introduced to cultivation by Forrest in 1917 and first flowered in 1924 and the possibility of spectacular advances is still with us.

❀ Primula sinensis

Another Chinese plant, but one for which Mr Li gives us no history, is *Primula sinensis.* It is a plant that has survived only in cultivation, no wild form being known. The species *P. rupestris* (syn. *P. calcicola*) coming from Shensi and Szechwan is evidently closely related. *P. sinensis* had certainly been in cultivation for many years, probably centuries, before it was introduced to Europe. In 1819 seeds and plants were brought from Canton, but the plants died and the seeds failed to germinate. The next year a successful importation was made and flowers were seen. They appear to have been a purplish-pink-flowered form with a yellow eye and were all thrum-eyed. Whether any seed was obtained is somewhat obscure as during the years from 1821 to 1826 a Mr Potts was sending seed regularly from China and it was from these importations that the present race originated. The first pin-eyed plant was noted in 1824. In 1827 the first white-flowered form was noted and fringed flowers appeared in 1833. By 1839 Anderson was offering double white and purple forms, presumably only semi-double as they were fertile. Breeding was also going on in Germany and by 1842 Bosse was offering nine varieties which included a double white. The colours available appear to have been purple, pink and white, but by 1846 a crimson form had appeared.

About the same time, fully double sterile flowers had been raised, but, since they had to be preserved by vegetative propagation, which is not easy with this species, they were not very popular. Around 1854 it would appear that Bosse raised a cultivar with a fern-like leaf, but this was lost. Between 1861 and 1865 the fern-leafed forms were raised again and this time preserved; and a large number of fertile semi-double forms were raised in colours of magenta, carmine, red, pink, white and blue. In the same period flaked flowers were noticed for the first time. Whether this feature was due to a virus is not certain, but it seems likely. By 1869 there were at least twenty well-defined cultivars in cultivation. A number of forms with large or small coloured eyes had been raised and there was one cultivar with variegated foliage. There was little notable advance, except for various leaf-form variations, until 1906 when a form with coral flowers appeared. In 1909 tetraploid forms were first observed; they appear to have occurred spontaneously, without any deliberate chemical inducement. In 1919 the so-called 'Harlequin' flowers with irregular petals and blotched flowers were introduced, but failed to commend themselves. In 1933 the famous cultivar 'Dazzler' with scarlet flowers was introduced. Shortly later 'Vivid' gave a similar colour to the stellata type of the plant.

At one time there were thought to be two species: *P. sinensis* with round flowers and *P. stellata* with the petals obviously separate giving a star-shaped flower. Both forms occurred in Potts's batches of seeds from China and are evidently different forms of the same species.

*Primula sinensis* has been the subject of great genetic investigation, culminating in D. de Winton and J. B. S. Haldane's 'The genetics of *Primula sinensis*' in the *Journal of Genetics* in 1933. According to this the flaked flowers are due to the recessive counterpart of the gene that spreads anthocyanin over the petals. The letter *E* designates this gene. On the other hand 'Harlequin' flowers are due to the recessive counterpart of the gene *H*, which makes petal colour and size uniform. In all, de Winton and Haldane found twenty-five alleles and two sets of three multiple alleles in the diploid form. There appear to be rather fewer in the tetraploids, but to compensate for this, the number of different combinations is naturally greater. Owing to the heterostyly of *Primula* spp., cross-fertilization is generally required and it must have been difficult to perpetuate some of the earlier cultivars. It was not until the 1880s that Suttons was able to sell seed of cultivars that could be guaranteed to breed true.

There are two other Chinese primulas, both of comparatively recent introduction, that have been bred as decorative pot plants in Europe, but not, apparently, by the Chinese. These are *P. obconica* and *P. malacoides*.

## ❀ *Chrysanthemum*

Although the introduction of the remontant form of *Rosa chinensis* has perhaps transformed European gardens more than any other single plant the Chrysanthemum must be regarded as of nearly equal importance. Here all the work had already been done before the plant reached Europe. Written evidence traces the cultivation of the Chrysanthemum back to at least 500 B.C. These earliest Chrysanthemums were all yellow in colour and apparently were forms of *Chrysanthemum indicum*. This is not, in itself, a very attractive plant. It has a head of small single yellow daisies and looks rather like a superior hawk-weed.

Presumably selection gave larger flowers and, since we are dealing with a member of the *Compositae*, flowers composed entirely of ray florets. It is not until the T'ang dynasty (A.D. 618-902) that we hear of the larger-flowered white Chrysanthemum and about the same time there is mention of a purple-flowered form. This, it is thought, must refer to *C. morifolium*, which is not dissimilar to *C. indicum* but with larger flowers which are white or purple in colour. While these plants were being cultivated, they must have become hybridized, as it is generally agreed that the cultivated Chrysanthemum, *C.* X *hortorum* arose originally from hybrids of the two species. It was recorded quite early on that new varieties could be raised by sowing seeds and these would be propagated vegetatively.

In A. W. Anderson's *How we got our flowers* (London, 1956), there is a vivid description of the Empress Hsu taking and caring for cuttings.

Around the end of the fourth century the Chrysanthemum was received in Japan and in 1797 was made the personal emblem of the Mikado. The emblem of the Japanese flag is a stylized sixteen-petalled single chrysanthemum.

The Chrysanthemum was encouraged to produce fresh varieties by grafting it on other plants, but we do not know if much success was attained in this way. The earliest Chinese monograph on the flower dates from the beginning of the twelfth century and lists thirty-five cultivars. At the end of the century the poet Fan Ch'eng-ta recorded seeing paintings of more than seventy different forms. When the famous *Kuang Ch'un-fang-p'u* (Enlarged thesaurus of botany) was published in 1708 nearly three hundred different cultivars were described. Nowadays the number of cultivars must be numbered in thousands. Although every variation was encouraged, it is perhaps safe to say that the incurved type was more popular in China, while the more informal bloom was encouraged in Japan.

The Chrysanthemum did not become established in European gardens until the end of the eighteenth century. They were first recorded in Holland in 1689, but the plants did not survive. There was also a yellow-flowered Chrysanthemum in the Chelsea Physic garden in 1764, but this, whatever it was, also was lost and the history of the Chrysanthemum in Europe really begins when Captain Blanchard of Marseilles brought into France in 1789 a white, a violet and a purple plant. Only the last one survived. Plants were obtained at Kew in 1790, but do not appear to have flowered until 1796. In this year the nurseryman Colville of Chelsea displayed large numbers. Imports of plants and seeds continued, one of the chief suppliers being James Reeves, the East India Company's resident tea buyer at Canton, from whom so many Chinese introductions have come. By 1830, seventy cultivars were known in England. In 1832 seed was produced for the first time from European plants (some authorities say 1827) and systematic breeding began. In 1838, John Salter, one of the first breeders of a white Fuchsia, who had a nursery at Versailles (he later moved to Hammersmith), collected all the cultivars then known in Europe and started breeding them. By 1842 there were at least 110 cvs. available in England. Raising Chrysanthemums also became a major industry in Jersey, where at one time there were said to be 4,000 different cultivars available. In 1846 Robert Fortune brought back the Chusan Daisy, a miniature form from which were bred the small-flowered pompon varieties. In 1861 Fortune brought back the larger-flowered reflexed cultivars from Japan, but they were not, at first, much liked. As most of the work was already done when the plants came to Europe, there has been remarkably little progress since that time. Between 1880 and 1890, a noted breeder, M. Delaux, worked on the development of cultivars that would flower early enough for them to be grown out of doors; and the Anemone-centred varieties, where the central disk florets are considerably enlarged, also seem to be the result of European breeding. Otherwise our only contribution has been hybridizing early flowering Chrysanthemums with *C. coreanum* (which may itself be a hybrid) to produce the many-flowered heads of the Korean strain. Indeed it is said that many of the best Chinese and Japanese cultivars have never been introduced to the west, owing to the fact that they are not considered to be commercially attractive. In recent years a few with spoon-shaped florets have been introduced *via* India, which suggests that further developments are possible. The popular cultivars with quilled petals known as 'Rayonnante' have not been developed since their original introduction.

Although we have little to congratulate ourselves on as far as the development of the flower has gone, we may congratulate ourselves on unravelling its probable parentage. With a basic haploid number of nine the *Chrysanthemum* spp. show polyploidy even in the wild state. Thus *C. indicum* is a tetraploid with $2n=36$ while *C. morifolium* is a hexaploid with $2n=54$. The cultivated clones have not $2n=45$ as one might expect, but varying numbers such as 52, 56 and 68. How these numbers developed is not clearly understood, but when a plant has been cultivated for two thousand years, it is not surprising that the more fertile mutations should be selected. One would expect the original hybrid *C. indicum* X *morifolium* to be sterile and any fertile hybrids must have been due either to apomixis, which is not found nowadays in *C.* X *hortorum*, or to an alteration in the original parent. It may well have been that a polyploid form of *C. indicum* had arisen in the course of its long cultivation and been preferred to the wild form. If it were a hexaploid ($2n=54$) the resultant hybrid with *C. morifolium* ($2n=54$) would be $2n=54$, which is what we actually find in some clones, while if it were octoploid ($2n=72$), we would get $2n=63$ in the hybrid. What we actually have among the clones that have, so far, been examined is 58 and 68.

## ❀ *China Aster*

Although countless wild flowers from China embellish our gardens, the only other two plants that have been developed in many ways are annuals. The China Aster, *Callistephus chinensis* is, in its typically wild state, a branched annual with violet female ray florets and yellow hermaphrodite disk florets. Seed was sent originally to Paris in 1728 and the resultant plants were all single and red, violet or white in colour. It was not long before double flowers were produced and other variations followed in course of time. Quilled petals were developed in Germany and a dwarf cultivar appeared about 1834. A recent variant is the pale yellow colour. Here again we may suppose that most of the work had been done before the European breeders started growing the plants and that the various mutants had already been isolated.

### ❀ *Chinese Pink*

The other popular annual is the Chinese Pink, *Dianthus chinensis*. Seeds of this were sent to Paris in 1705 by the missionary Bignon and double forms appeared soon afterwards. However their most spectacular development seems to have been due to the Japanese, rather than the Chinese. In the later nineteenth century, a Mr Heddewig, a gardener at St Petersburg, imported seed from Japan and the offspring of these became known as the 'Heddewigii' strain, which are gaily coloured with fringed petals. A Frenchman, M. Chabaud, succeeded about 1870 in crossing *D. chinensis* with the Carnation and thereby raised the race of large-flowered annual Carnations that are known as Chabaud or Marguerite Carnations. This, like the Rose, shows an interesting mixture of Chinese and European cultivation. As we shall see when we come to discuss the plant the European Carnation has a long, even though rather obscure, history. Of course, compared to the Chinese, any European garden plant is comparatively recent, but only the Rose and, perhaps, the Tulip has a longer period of cultivation. One wonders if the cross could be pursued further to include the Pink, *Dianthus plumarius*, which might bring back a perennial plant with the colours of the Chabaud Carnations.

Since no country has so long a history of gardening as China, so no country has contributed so much to the gardens of all temperate climes.

CHAPTER IV

# Plants developed in Turkey and the Middle East

*Anemone – Ranunculus – Hyacinth – Lilac – Tulip*

Most of the plants described in this chapter reached Western Europe from Constantinople; the majority owing to Ogier Ghiselin de Busbecq, the Viennese ambassador there. However the plants were probably generally cultivated in the Middle East and in Persia and should not be regarded as exclusively Turkish. Some plants are said to have been brought back by the Crusaders, but these traditions lack any particular verification. There appears to be very little contemporary literature on the products of Turkish gardening, or, if there is, it has not been translated into any language known to me. As might be expected, most of the cultivated plants were selected forms of local wild flowers.

## ❀ *Anemone*

Double forms of *Anemone coronaria* and *A. hortensis pavonina* were very popular and have continued to this day, but were even more popular in the eighteenth and early nineteenth centuries. Miller wrote that they had been 'so greatly improved by culture as to render them some of the chief ornaments of our gardens in the spring. The principal colours of these flowers are red, white, purple, and blue, and some are finely variegated with red, white and purple. There are many intermediate shades of these colours; the flowers are large and very double, and, when properly managed, are extremely beautiful.' Miller's 'proper' management was extremely complicated, but must have been very successful. A mixture of two parts sandy loam (with turf) and one third rotten cow dung was seasoned for twelve months and this mixture was the basic soil; it was laid 30 inches thick on 4 or 5 inches of spent manure. In wet situations the surface of the bed should be raised from 6 to 8 inches above the general level, while in dry situations 3 inches is sufficient. With slight variations in the soil mixture, this is the treatment that Miller advises for all bulbous plants and it is to be expected that superlative results would obtain.

It is difficult to understand why the anemones had to be introduced from Turkey, as *A. coronaria* is wild in Italy. There is indeed a legend that the scarlet form appeared in the Campo Santo at Pisa among ballast from the Holy Land, brought back by ships which had been delivering crusaders. However, since it was in Italy anyway, it scarcely seems necessary. It is true that the scarlet form is commoner in the east, while the violet form is more prevalent in the west, but both are to be found in Italy. It is also seen in southern France, apparently growing spontaneously. It seems to have been the Turks who selected the double forms, which were regarded as most attractive. Since the sixteenth century there seems to have been little development. It was never sufficiently regular to become a 'florist's flower' and so did not get the selection and vegetative propagation of special forms that characterize other plants of Turkish origin, such as Ranunculus, Hyacinth and Tulip. Nowadays we are offered 'strains' and a few named cultivars, which, presumably, will breed true from seed.

*A. coronaria* can be hybridized with the various subspecies of *A. hortensis*, but nothing very startling seems to result. It is possible that the 'St Bavo' strain, though deriving principally from *A. hortensis*, may contain some *A. coronaria* blood.

### ❀ *Ranunculus*

*Ranunculus asiaticus*, another native of the Middle East (also found in Crete and Libya) is variable in nature; scarlet, yellow, orange and white forms are to be found and it seems prone to doubling. It had already been cultivated for a long time when it reached western Europe in 1596 and was to be very much further developed in the eighteenth and early nineteenth centuries. In 1835, for example, no less than 140 named cultivars were exhibited at the Cambridge Florists' Society and in 1792 Maddock had more than 800 named cultivars. Of course many of these must have been hard to distinguish. Incidentally there is a tradition that Louis IX brought it to his mother Blanche of Castille from the Crusades, but, even if this is true, it did not enter into general cultivation until the end of the sixteenth century. In 1629 Parkinson enumerated eight varieties. In the nineteenth century the plants were shown 'in classes' listed as follows: white ground spotted, yellow ground spotted, white ground edged, yellow ground edged, dark purple, rose and pink, light purple and grey, orange, black, buff, red and white striped, olive, white, crimson, yellow, scarlet, coffee colour, red and yellow striped, shaded white, and mottled. Contemporary illustrations depict very symmetrically disposed, fully double flowers of delicate colouring with the petals tipped with another colour (these were, presumably, the white and yellow ground spotted). The mention of olive, grey and coffee colours is interesting, as they appear now to be lost and it is not easy to see how they occurred. The black was of course a very deep purple, but purples are not commonly found among the wild forms and cultivation and selection must have produced some interesting mutations. The striped forms sound as though they might be due to virus infection. As *R. asiaticus* is not reliably hardy in this country it is interesting that they should have been so popular here. Admittedly it is as easy to store Ranunculus corms as to store Dahlias, but their cultivation is slightly more difficult. According to Miller's *Gardener's Dictionary* it was common practice to plant the corms in the autumn in the eighteenth century, but by the early nineteenth century February planting was preferred. Miller's recommended soil mixture was excessively elaborate even for him and had to be prepared for two years before it was used. It was basically made of rotted turves, manure and sand and Miller claimed that with it he was able to produce fifty flowers from a single plant. The named cultivars were propagated by offsets which, according to Miller, 'they generally produce in great plenty, if planted in good soil'. This is more than they seem to do nowadays and one reason for the proliferation of cultivars may have been the difficulty in providing sufficient of any one cultivar. The cultivation of 'florist's' or 'mechanick's' flowers seems to have been more or less confined to Northern Europe, principally England and Belgium and, as a result, once Ranunculus ceased to be fashionable, the various forms became extinct. Had they been popular in Mediterranean countries, they might have been sought out in the equivalent of cottage gardens, much as the older roses have been rediscovered in our day. As it is, however, almost all these fine forms have been lost and the olive, black, coffee and grey colours have vanished from any strain of *R. asiaticus* now in cultivation. Whether there is any commercial inducement I do not know, but it would seem that here was a field for the plant breeder. He could re-create these vanished beauties. It must be entirely a matter of selection, as no one has ever suggested that any other species was ever hybridized in. The history of the Ranunculus shows how tenuous the existence of man-made flowers is. If once they get lost, there is all the work to do again. It seems that this occurs most frequently with tender plants. During the latter half of the nineteenth century there was much hybridizing done among the various Cape Heaths. These are now lost. Similarly only a very few of the so-called Javanicum Rhododendrons now survive.

### ❁ *Hyacinth*

On the other hand the Hyacinth, another product of Turkish gardening is still very much with us, although in fewer varieties than in the olden days. The species, *Hyacinthus orientalis*, is common throughout the Middle East in rather high situations. It is usually pale blue in colour with rather thin leaves. The popular white Roman Hyacinth looks like the wild form so far as its flowers are concerned, but has rather wider leaves. Originally only four colour forms were known, but by 1725 no fewer than two thousand named cultivars were listed. Modern forms are very much larger than the wild type and appear to have collected many extra chromosomes which do not relate to the haploid number. This is 8 and the wild form has $2n=16$, but the cultivated plants have numbers ranging from 19 to 30. According to Crane and Lawrence the first triploid Hyacinth to appear was 'Grand Maître' in 1870. The matter is not too clear as they illustrate a pollen grain of a Hyacinth showing 12 chromosomes, which seems to suggest that the triploid acts as a diploid with a haploid count of 12. It is to be noted that, unlike the products of Chinese gardening, most of the Turkish plants have been considerably developed since their introduction to western Europe. The four original forms were a pink, a blue and a single and a double white. The colour range has been extended to include yellow, deep crimson and purple. At one time the double flowers were the most prized; these were literally double, with two corollas, and an occasional triple form was bred. Subsequently the double forms waned in popularity and most of them have been lost. Of late years a few have been reintroduced. Polyploidy is said to be not uncommon among plants reproduced vegetatively, either by grafting or cuttings, or by means of offshoots from bulbs, corms and tubers. If Crane and Lawrence are correct in saying that the first triploid did not occur until 1870, the former Hyacinth must have been due to simple gigantism. Otherwise the development of the flower seems to be due entirely to selection and the crossing of selected forms. The breeding of Hyacinths has been almost entirely a Dutch accomplishment, although there was at one time a thriving Hyacinth industry around Berlin.

### ❀ *Lilac*

Lilac, *Syringa vulgaris,* also came from Turkey. The plant is native to Rumania and Bulgaria, but was not found in a wild state until 1828. It was known in three colours, white, blue and purple and subsequently a double white form was found. Later still the double blue form was used as the basis for his hybridization by Victor Lemoine in the 1870s. Besides the various forms of *S. vulgaris* Lemoine also used pollen from the Chinese *S. oblata,* which had been brought to Europe by Fortune in 1856. The modern Lilac is thus a comparatively recent plant, but the increase has been principally in size and the number of doubles. It is true that there is a pale yellow cultivar 'Primrose', but otherwise the colours are the same as they were in the seventeenth century: white, blue and purple. There are one or two cultivars that have purple or mauve flowers tipped with white. These, according to Dr Haagedoorn;* may well be periclinal chimeras: a phenomenon in which the outermost layer of cells is of a different nature to those underneath. In the case of Lilac it is assumed that the outer layer is white, while the inner layers are purple. As the cells are only two deep at the edges, the white colour predominates, while the petals are thicker in the centre and the purple shows through the single layers of white cells.

### ❀ *Tulip*

There are many bulbous and tuberous plants that we owe to the Turks, such as the Crown Imperial (*Fritillaria imperialis*) a native of Persia and the western Himalayas that was much cultivated in Turkey, where it has become naturalized; and *Iris susiana,* a native of Lebanon and Syria, the wild forms of which seems to have acquired the names of *I. basaltica* and *I. sofarana*; but these do not seem to have been further developed. One bulbous plant that has been very much developed is the most important plant of Turkish origin: the Tulip. And yet

* *op. cit.*

it is not known from what species it was first bred. It seems to have been that excellent Ghiselin de Busbecq who first saw it between Adrianople and Istanbul in 1554. He obtained some bulbs which he grew himself and also gave to such people as the rich Fuggers (in whose garden at Augsburg Conrad Gesner saw it in 1559) and to botanists such as Clusius. Bulbs were stolen from him and seem to have formed the nucleus from which the great Dutch industry started.

We have, however, to ask ourselves whence the Tulip came. Now there is no recognized Tulip species that can be regarded as the ancestor of the garden Tulip. It is distinguished from all other species in that varieties with flowers having both a yellow and a white ground colour are to be found. But in the section *Gesnerianae*, to which the garden Tulips are most akin, no species with a white ground is known; all wild species of this family have a yellow ground. Moreover, experimental breeding has shown* that the yellow ground colour is a simple mendelian character, recessive to white ground. Sir Daniel Hall† suggests that a mutation from a yellow to a white ground arose in a cultivated specimen of some original Gesnerian species and this, by cross-breeding with the type, gave rise to the cultivated race. He adds that 'mutations to a dominant are not the general rule', but that 'they are not infrequent in other genera that have been intensively studied'. Such a statement from someone with the authority of the late Sir Daniel Hall must be accepted, but I could wish he had given some examples of this phenomenon. I must confess that I am not convinced by it. The wild forms of the Gesnerianae, with two exceptions, are scarlet-flowered with a black blotch at the base and a yellow margin. One of the exceptions is *Tulipa galatica,* which has pure yellow flowers and a number of fragments in addition to the usual 24 chromosomes. It is only recorded from two districts in Turkey and would appear to be derived from *T. armena.* This is an interesting plant because of its peculiar cytology, but probably not of great importance in trying to solve the problem of the

* Sir Alfred Daniel Hall, *The Genus Tulipa* (London, 1940).

† *Ibid.*

origin of the garden Tulip. The other exception, *T. marjolettii*, will be discussed shortly.

It is, of course, possible that the original ancestor of the white ground Tulip was collected to such an extent that it has become extinct in the wild. We have seen the same thing happening with *Primula sinensis*, very nearly with *Paeonia suffruticosa* and altogether with Ginkgo and *Malus spectabilis*. Both Omar Khayyam and Hafiz refer to Tulips in their poems, which suggests that they were cultivated in the twelfth and thirteenth centuries, although there is no representation in either Persian paintings or tiles before the sixteenth century. Whether Omar Khayyam's Tulip was a hybrid or one of the wild species cannot be determined, but it may be significant that *T. clusiana* was known to the early botanists as the Persian Tulip. *T. clusiana* is unique among Tulips in being a pentaploid. It appears to have been derived from the tetraploid *T. chitralensis*, which in turn has arisen from the diploid *T. aitchisonii*. Both the diploid and tetraploid plants come from Kashmir and the Chitral district and may well have been imported into Persia at the same time as the yellow form (var. *chitralensis*) of the Crown Imperial. Certainly *T. clusiana* seems to have reached western Europe from Turkey and has become naturalized in some Mediterranean places. Although one would expect a pentaploid to be sterile, it has been known to set seed spontaneously and this seed reproduces the parent form. It may well be that this Tulip is the one referred to by Omar Khayyam and Hafiz and our common garden Tulip would then be a product of the late fifteenth and early sixteenth centuries.

Towards the beginning of the nineteenth century botanists began to discover Gesnerian Tulips apparently growing wild round Florence and in Savoy. These were given specific names, but are now regarded as forms of early garden Tulips that have escaped from cultivation and are known collectively as the Neo-Tulipae. It seems probable that the plants round Florence must be escapes, but the Savoy plants are in a different category. Tulips were found in two localities – St Jean de Maurienne and Aime. Neither is at all the sort of place where one would expect to find garden Tulips naturalizing themselves. They are both mountainous districts; St Jean de Maurienne is 565 metres above sea level while Aime is 1,355 metres. What is probably significant is that both lie on roads leading to major passes through the alps between France and Italy. St Jean de Maurienne is on the road leading over the Mont Cenis pass, while Aime is on the road leading over the Little St Bernard. It would seem not unreasonable to assume that the Tulips must be the result of some jetsam from pack animals bringing bulbs overland. All the Neo-Tulipae are some shade of red with two exceptions. *T. billietiana* is a pure yellow, while *T. marjolettii* (syn. *T. perrieri*) is white (or rather cream) with a pink flush as the flower ages. Now this is not only the sole member of the Neo-Tulipae that is white, it is the only white Gesnerian Tulip ever found growing apparently wild. It is an attractive plant that is still in cultivation and it does not seem to me to be straining credibility too far to suggest that it must be very near one of the ancestors of the garden Tulip. It is one of the two Neo-Tulipae reported from Aime and the situation is sufficiently isolated to prevent much pollination from cultivated plants. It is feasible that in this isolated place some Tulips were jettisoned and naturalized themselves and possibly in the course of time segregated themselves out to something near their original ancestors. The other Aime plant, *T. aximensis,* is deep red with a yellow-margined green blotch, close enough to most of the Gesnerianae. Incidentally neither of the Aime Tulips was discovered before 1894. I would suggest that someone with a longer expectation of life than I have, should try crossing *T. marjolettii* with some of the Turkish and Persian Gesnerian Tulips and see if the results resemble the garden Tulip. As one would want to continue for at least three generations, it is to be supposed that a period of from fifteen to eighteen years would be needed to test this theory. Anyway for what it is worth I suggest that one of the ancestors of our garden Tulip was a white Tulip, not unlike *T. marjolettii,* which was only found in a restricted area and which was collected to extinction. As Hilaire Belloc so pithily puts it:

Oh let us never, never doubt
What nobody is sure about.

Until recently no other species were bred into the original Turkish hybrid, with the possible exception of *T. suaveolens*, a dwarf early-flowering Tulip from Southern Russia, which may have entered into the Duc van Thol strain. I suspect, however, that there may be another missing species which would account for the forms with pointed petals. This is to be found in some of the Neo-Tulipae and was bred into the so-called *T. retroflexa* by a certain V. van der Winne whose collection was sold in 1863. It is from this that the so-called lily-flowered Tulips derive. Even more extraordinary is the cultivar known as *acuminata*. This is a small plant with very long thin flowers with the petals excessively elongated; they sometimes reach a length of 6 inches but are barely an inch across at the base and end in a threadlike portion. Of course in plants that have been interbred so much one must expect recessive characters to emerge and this attenuation of the petals may be one of them. Reflexed petals were at one time always associated with weak stems and it is only comparatively recently that lily-flowered Tulips with strong stems have been available.

From the time of their introduction until quite late in the nineteenth century, the ideal Tulip was 'broken'. Self-coloured Tulips were of no account except as 'breeders' of Bizarres and Bybloemens. We now know that this effect is caused by a virus, which is transmitted by aphids, but this knowledge has been acquired too late to be of any use to plant breeders. The broken Tulip is no longer admired. Illustrations of admired Tulips in the 1830s show that, even at that time they were very different from the kind of flower that is to be seen in Dutch flower paintings. The ground colour is emphasized and the feathering is confined to the tips of the petals, so as to give the flower the appearance of a modern cultivar.

The future of the Tulip is difficult to foresee. If we breed in *T. praestans*, it should be possible to have several flowers to the stem and *T. greigii* will contribute leaves striped with purple.

No new colour breaks appear very likely. Perfume has already been bred into many cultivars, but no attention has been drawn to this desirable attribute. Whether a completely new set of garden Tulips could be bred from the smaller flowered Eriostemones is difficult to say. They will not hybridize with the other group, the Leiostemones, which includes the garden Tulips, but they might well hybridize among themselves. It is doubtful whether anything very desirable from the commercial point of view would emerge. It might also be possible to prolong the flowering season by breeding in *T. sprengeri*, but this species does not seem to hybridize at all easily. Recent developments have tended to ignore the garden Tulips as at present constituted and start afresh with recently introduced species such as *T. greigii, T. eichleri, T. fosteriana* etc. A number of delightful cultivars from *T. kaufmanniana* have also been bred and there are now hybrids between these and both *T. fosteriana* and *T. greigii. T. fosteriana* gives flowers of great size, but the hybrids have not yet been so successful as the more favourite Gesnerianas. However they have not yet had time to reach beyond the second generation and the future may have surprises in store.

Finally, we come to the Rose. The question of which Roses had their origin in the Middle East is by no means easy to determine, as their cultivation goes back to the earliest civilizations. As already mentioned, in connection with Roses originating in China, it seems more convenient to deal with all the Rose species together, including those that came to Europe from the Middle East.

CHAPTER V

# The Rose

Since China and the Middle East have bred many of the ancestors of the modern Rose, this seems a suitable place to consider the development of one of the most popular of all flowers.

The genus *Rosa* is found throughout the north temperate zone. The largest number of species seems to be in China, but there are no countries in this area without some species.

The development of the Rose is a unique example of the fusion of two garden cultures: that of China and that of Europe. It might be more accurate to say of Eastern Asia and Western Asia, as it is thought that many of the European roses may have originated in Persia and in Asia Minor. However, whatever the origin of the European Roses may have been, they had been developed in Europe for at least three centuries before they were bred with the products of China.

Three ancient Roses from which are descended many modern hybrids are *Rosa gallica*, *R. phoenicea* and *R. moschata*. *R. gallica* was known to the Persians in the twelfth century B.C. and may have been brought to Southern Europe at the time of the Persian invasion of the West. The French have been making pot-pourri and conserves from dried petals of *R. gallica officinalis* since the fourteenth century, and the flower must have been familiar in this country by the fifteenth century, for *R. gallica* is supposed to have been the Red Rose of Lancaster. One would have expected it to have been the first Rose to be cultivated in Europe, but in fact the earliest Rose to be depicted, on one of the Knossos frescoes, is the Damascene Rose, a hybrid of *R. gallica* with *R. phoenicea*. The latter rose, a member of the group known as *Synstylae*, has a sprawling habit and corymbs of very small white flowers. It much resembles *R. multiflora*, a parent of many 'Rambler' roses, and there seems no reason to suppose that it was consciously used as a parent in the hybrid.* It is possible that *R. phoenicea* would contribute larger corymbs than are to be expected in *R. gallica* itself and it would also inhibit the suckering habit that characterizes *R. gallica* and thus make a more compact plant. Another, less desirable quality it has contributed is rather fierce thorns: *R. gallica* itself is nearly thornless and the prickles, when present, are very soft. The hybrid *R. damascena* has numerous prickles.

The *gallica* X *phoenicea* hybrid seems to have been taken to Abyssinia by some early Christian and is found growing around early church sites. It has therefore been known, quite incorrectly botanically, as *Rosa sancta*. It also seems that cultivars of the hybrid were brought back to Western Europe, particularly France, by the Crusaders. The *Roman de la Rose*, written about 1260, refers to roses from the lands of the Saracens. Indeed many of the so-called Gallica Roses are not cultivars of *R. gallica*, but of *R. damascena*, the *gallica* X *phoenicea* hybrid.

*R.* X *bifera* (or *R. damascena bifera*), the Autumn Damask Rose, was thought by Dr Hurst to be a hybrid between *R. gallica* and *R. moschata*. It is referred to by Herodotus and is the twice-flowering Rose of Paestum mentioned by Virgil in the Georgics; it is also depicted in the frescoes of Pompeii. The hybrid which, according to Herodotus, had sixty petals, was said to have been introduced into Greece, initially perhaps to Samos, by King Midas, who had the flower in his gardens in Macedonia. It has been assumed that this was a double-flowered cultivar of *R.* X *bifera*, and it may well have been an ancestor of the roses that are so extensively grown in Bulgaria nowadays for the extraction of attar. By no means all the Damask Roses flower twice a year, but the cultivar known as 'Quatre-saisons' and its white-crested sport, the 'Perpetual White Moss' certainly do. Though known

* This is assuming that Dr C. C. Hurst's diagnosis of its parentage is correct. All Dr Hurst's suggestions are very probable, but they are entirely notional. No one has attempted to reproduce his suggested hybrids. See his papers on the subject in the R.H.S. *Journal* for 1941.

to the Greeks and Romans, *R.* X *bifera* does not seem to have reached beyond Italy until brought thence by King Henry VIII's physician, Dr Linacre, shortly before his death in 1524.

Owing to the propensities of roses to hybridize, we may assume that the Autumn Damask is a chance hybrid, which would be perpetuated for its valuable properties. *R. moschata*, another of the *Synstylae*, is a climbing Rose with a late-flowering season; in this country it comes into flower at the end of August and it seems to contribute this late-flowering habit to any hybrid in which it is involved. The so-called Hybrid Musk Roses of today have the most infinitesimal portion of *R. moschata* in their ancestry, but they are all noted for their superlative display in late autumn. It is, one would have thought, highly unlikely that *R. moschata* could hybridize with *R. gallica*, which usually flowers two months earlier, and the hybrid must have been due to some unusual coincidence of either *R. gallica* having some late flowers or *R. moschata* some early ones.

The native habitat of *R. moschata* is not known, but it was cultivated in Persia and the Mohammedans seem to have taken it with them wherever they went, so that the early writers gave such places as Spain, Morocco and Madeira as its native land. In Redouté's *Les Roses* Thory says that it is a native of Indostan. We know that some Indian plants were cultivated in Persia, so such a habitat is not unlikely, but no one seems to have found it there recently. Thomas Rivers in the *Rose Amateur's Guide* (1843) speaks of plants being raised from a 'Chinese Rose Tree' at Isfahan. The seedlings were all common Musk Roses. The references to Spain and Barbary suggest that the Arabs probably took it to Spain and North Africa; but we can say for certain that it is not found wild in either Spain or Madeira. In North Africa it seems to be naturalized in one or two places, but Persia would seem to be its more probable homeland. Although at one time very popular, it is now extremely rare in cultivation. A fascinating account of its rediscovery is to be found in Graham Thomas's *Climbing Roses, Old and New* (London, 1965).

Another ancestor of the garden roses is the White Rose,

*R.* X *alba,* which is thought to be a cross between *R. damascena* and a white Rose of the *Caninae* section. Whether this rose, too, had an Eastern Mediterranean origin is by no means certain. It was certainly known to the Romans, and is believed to have been the White Rose of York. Owing to the curious pairing of chromosomes in *R. canina,* we can be sure that it was the female parent of *R.* X *alba.* The haploid count of the genus is 7, but there are numerous polyploids among the species and the *Caninae,* to which our common Dog Rose belongs, have 35 chromosomes. Seed is sometimes set apomictically, but sexual pairing can also occur. When it does, the fertile pollen grain carries 7 chromosomes, while the egg-cell carries 28 (so that the plant resulting from normal fertilization will carry 35 chromosomes). The Summer Damask has 14 chromosomes in both male and female germ cells. *R.* X *alba* has 42 chromosomes in its somatic cells, which at meiosis divide into 28 in the female germ cells and 14 in the male.

It is somewhat ironic that, although we consider the Rose to be particularly European, when it is not Chinese, only one of the main groups of roses seems to have been raised in Europe. This is the *centifolia* or Cabbage Rose, raised about 1710, probably by crossing an Alba with an Autumn Damask, and therefore containing the genes of *RR. gallica, phoenicea, moschata* and *canina.* Roses have very large numbers of stamens and many of these can be converted into petals without the resultant plants being sterile, and this seems to happen quite frequently in the wild. The doubling appears to be a dominant characteristic and is to be found in $F^1$ hybrids and in selfed plants. The original Cabbage Rose, *R.* X *centifolia,* was exceptional in being fully double and therefore completely sterile. However, a number of bud-sports occurred, including the Moss Rose, so there are quite a large number of centifolia cultivars. Obviously there is no room for further development in this section, except by the emergence of more bud-sports, an unlikely occurrence now that the plant is little grown.

The older garden roses discussed so far are quite complicated enough as they stand, but their nomenclature has made matters even more difficult. The officinal form of *R. gallica* was greatly cultivated around Provins in the department of Seine-et-Marne to the south-east of Paris. As a result *R. gallica* became known as the Provins Rose. When *R. centifolia* was bred, it became known as the Provence Rose. The possibilities for confusion have been fully exploited. Similarly there is a confusion between *R. damascena,* the Damascene or Summer Damask Rose which is *R. gallica* X *phoenicea,* and *R.* X *bifera,* the Autumn Damask, sometimes just referred to as the Damask Rose, which is *R. gallica* X *moschata.* Difficulties over Rose nomenclature have continued, in some cases even up to the present day.

Towards the beginning of the nineteenth century, several hybrids of *R. arvensis* were developed at Loudoun Castle in Ayrshire and became known as Ayrshire Roses. *R. arvensis* is a British native and it is probably this Rose and not *R. moschata* that Shakespeare and Keats refer to as the Musk Rose. However, according to Sabine, a secretary of the London Horticultural Society, the seeds of *R. arvensis* that gave rise to the Ayrshire hybrids were sent from North America. It is hazarded that they may have been hybridized with *R. sempervirens,* a rather tender member of the *Synstylae,* which is nearly evergreen and which, in its turn, was a parent of two fine French-bred ramblers, 'Adélaide d'Orlïans' and 'Félicité et Perpetue' raised by M. Jacques at Neuilly between 1824 and 1832. The Ayrshire Roses had double white flowers and were very vigorous, but do not seem to have been developed any further.

The Burnet or Scots Rose was much cultivated in the early nineteenth century. Sweet's *Hortus Britannicus* of 1839 lists 177 cultivars of *R. spinosissima,* although Loudon, in his volume with the same title also published in 1839, lists only 26. Of these 16 seem to be in cultivation still. Many varieties grown under this name are now thought to be derived from *R. pimpinellifolia,* a nearly related species. In his book Sweet gives a list of names, whereas Loudon confines himself to descriptions such as Double Blush, Double Crimson, Double dark Marbled, Double light Marbled, Large double Yellow, Small double Yellow, etc. The interest seems to have been started by a

Scottish nurseryman Robert Browne, who collected wild forms and, presumably, improved them by sowing seeds and selecting the best forms. The Burnet Rose was also crossed with the Austrian Briar, the yellow *R. foetida,* and gave rise to several yellow Burnet Roses of which 'Harrison's Yellow' and 'Williams's Double Yellow' are still in cultivation. These must have been the first reliably hardy yellow roses to be presented to British growers. *R. foetida* itself is rather delicate, while the other yellow Rose known at that time, the sulphur yellow *R. Hemisphaerica* which like *R. foetida* originated in Western Asia, is very unlikely to open its flowers in Britain.

Some time in the mid 1830s *R. pimpinellifolia* appears to have been chance pollinated by either a China or an Autumn Damask and the resultant seedling was put on the market by the famous Lee of the Vineyard Nursery, Hammersmith, under the name of 'Stanwell Perpetual' (Stanwell, in Essex, was where the original seedling occurred). This has its main display of double blush-pink flowers at midsummer but continues to produce smaller amounts of flowers until the frosts come. With the introduction of remontant roses, the spinosissimas fell from popularity, but have recently been used again, notably, as we shall see, by Herr Kordes.

The Sweet Briar, *R. rubiginosa,* has always been a favourite for its attractively coloured flowers and its perfumed leaves, and variants were eagerly sought. Sweet lists nineteen forms, including a double mossy form. The cultivar known as 'Clementine' may be the same as 'Janet's Pride', although this latter was not introduced under that name until 1892. 'Clementine' was known in 1830. If it is the same as 'Janet's Pride' it is slightly double with cerise petals and a white eye. About 1890 Lord Penzance started a series of crosses between the Sweet Briar, various Hybrid Perpetual and Bourbon Roses and also *R. foetida.* The resultant hybrids, known as the Penzance Roses, are, with the exception of the two *foetida* crosses, rather similar in appearance, being various shades of red with white centres, but have aromatic foliage, although this is not so strong as in the species itself. Only Herr Kordes seems to have used *R. rubiginosa* again for other hybrids and its full potentials are still to be exploited.

The above-mentioned cover the majority of garden roses grown before Chinese roses were first used by breeders. In addition, a double form of *R. cinnamonea* was grown, and *R. moschata* and *R. sempervirens* were in cultivation, as well as the yellow *R. foetida* and copper-coloured *R. foetida bicolor*. The introduction of plants from China transformed the Rose completely from a once-flowering shrub to the remontant dwarf Hybrid Tea.

The wild form of *R. chinensis* is a climbing plant with a single flowering season. However it is well known that perpetual flowering roses will sometimes produce forms that climb and only flower once and the opposite phenomenon can also occur. Indeed it appears to be a classic example of the emergence of a recessive gene. If the gene that causes the development of climbing shoots is suppressed, the recessive factor gives rise to flowering shoots, with the result that instead of a tall climbing plant with one season of flowering we get a dwarf shrub that flowers continuously. This phenomenon produced the remontant form of *Rosa chinensis* which seems to have arrived in Europe considerably earlier than other Chinese plants. The plant had been known in Chinese gardens since the tenth century if not earlier. The Rose has always been a favourite garden flower and it must have been exported, possibly via Persia, a centre of rose culture, to Europe where it probably arrived during the fifteenth century. It is depicted by Bronzino in a painting that is assigned to the year 1529 and it would seem that Italy was the country where it was most cultivated. In 1578 Montaigne was shown a rose at Ferrara, which, he was told, flowered all the year round. After this date there seems to be a silence on the subject until we reach the eighteenth century. In Gronovius's herbarium is a specimen of a crimson China Rose dated 1733. In 1751 Peter Osbeck, a pupil of Linnaeus, made a botanical voyage to China and brought back several specimens of what was then termed *Rosa indica*. There are six specimens in Linnaeus's own herbarium, one of which, according to Dr C. C. Hurst,* is a hybrid between *R. chinensis* and *R. multiflora*.

Apparently Miller was cultivating the pink China Rose at Chelsea in 1759. However he does not appear to mention it in his famous *Gardener's Dictionary* although there is a reference to the 'Monthly Rose' which he assumes is a form of the Damask, but which might refer to this plant. There is also, in the French edition, published in 1785, a reference to a deep red China Rose being cultivated in England. This seems to have been lost. Miller's reference to the 'Monthly Rose' may refer to the so-called Autumn Damask.

An interesting fact brought out by Dr Hurst is that the particular mutation that causes the perpetual flowering of *Rosa chinensis* has been observed in other Chinese roses and observed nowhere else. It would seem probable that the same mutation must have occurred at an early period in the Tea Rose, *R. gigantea*. This must have been either accidentally or deliberately hybridized with *R. chinensis* and it is to the importation of the resultant hybrids that the evolution of all the modern roses is due. The first of these hybrids known to have arrived in Europe is Parson's Pink China which is said to have been introduced by Sir Joseph Banks in 1789 and first flowered in Mr Parson's garden at Rickmansworth in 1793. It was introduced to commerce as the Pale China Rose. In 1805 a miniature form, the Fairy Rose (X *lawranceana*) appeared.

In 1809 came another of these Chinese hybrids, which was one of Sir Abraham Hume's importations. This was Hume's Blush Tea-scented China. It created such a sensation that special arrangements had to be made by the British and French Boards of Admiralty to arrange the safe-conduct of plants for the Empress Josephine at Malmaison. Although the plant has been lost to cultivation, contemporary illustrations show it to be nearer to *R. gigantea* than *R. chinensis*. Sweet gave this rose the specific name of *R. odorata*, which must now be regarded as invalid. Even more striking, apparently, for the plant has again

* In papers in the R.H.S. *Journal* for 1941, reprinted in Graham Thomas's *The Old Shrub Roses*, London, 1955.

been lost, was Parks's Yellow Tea-scented China which was brought from China by Mr Parks for the Horticultural Society in 1824. It was imported into France by the famous rose breeder Hardy in 1825 and it was in that country that it was most generally used for breeding. It is by no means impossible that plants of these two lost forms may still exist in Yunnan, where many of these old products of Chinese gardening were recently found to have survived. In its way *R. gigantea* is of equal importance to *R. chinensis*. The latter species gave the remontant character, but *R. gigantea* provided the typical pointed centre of the Tea and Hybrid Tea Roses. *R. gigantea* is rather tender and imparted its tenderness to the Tea Roses. It is rare in cultivation, if, indeed, it is in cultivation at all, but the climbing Roses known as 'La Folette' and 'Belle Portugaise' are very similar in appearance, except for the petal colour, which is yellow in *R. gigantea*.

The least important of the four stud Chinas, to use Dr Hurst's splendid term, was in fact the first to be introduced to this country: the *R. chinensis* cultivar Slater's Crimson China. Since it was a triploid, the pollen was of little use, but it did bring a completely new colour into the Rose when it arrived here in 1792. Indeed it may have been known even earlier in Italy and may have been the parent of the 'Portland Rose' that was introduced from Italy by the Duchess of Portland and given the name of *Rosa paestana*, owing to the tradition of its having come from Paestum. It was given its English name by the French nurseryman Dupont, who obtained it from England in 1809. From this plant came the 'Rose du roi' which is thought to be the first of the Hybrid Perpetuals and to have been raised by crossing a Portland Rose with a Hybrid China (i.e. a hybrid between *R. chinensis* and a European cultivar).

It is interesting to note that the first successful developments of the China Rose occurred not in Europe at all, but in the United States. Around 1802 an amateur, John Champneys, fertilized *Rosa moschata* with the pollen of Parson's Pink China (*R. chinensis* X *R. gigantea*) and produced a climber with semi-double pink flowers. This became known as Champneys's Pink Cluster. John Champneys lived at Charleston, South

Carolina where there also lived a French nurseryman, Philippe Noisette, who indeed had sold Champneys the China Rose parent. Noisette sowed seeds of Champneys's Pink Cluster and, as a result of Mendelian segregation, some of the resultant plants had the remontant habit of the China Rose. These plants were less vigorous than their parent, but had dense clusters of pale pink flowers that were produced continuously from June until the frosts came. Philippe sent this plant to his brother in Paris, Louis Noisette, and it was distributed to commerce in 1819. It was known as the Noisette Rose.

Between 1819 and 1830, a crimson flower was introduced, presumably through Slater's Crimson and the Noisettes were also hybridized with *R. multiflora* and *R. sempervirens*. These were not, of course, remontant. Towards 1830 nurserymen began to cross the Noisette with Parks's Yellow in the attempt to raise a yellow Noisette. In 1830 two successful hybrids were introduced. These were 'Lamarque' and 'Desprez a fleur jaune'. The latter is creamy-orange, rather than yellow and is still a favourite climber; 'Lamarque' is a lemon-yellow fading to white. In 1833 'Smith's Yellow' was really the first yellow dwarf Tea Rose. The other Noisettes continued as climbers. 'Lamarque' when selfed gave rise to 'Celine Forestier' (1842) with pale yellow rather flat flowers. 'Jaune Desprez' when selfed gave rise to 'Cloth of Gold' (1843) and 'Solfaterre' with rather larger sulphur flowers. From 'Cloth of Gold' came perhaps the most famous yellow Rose of all, 'Maréchal Niel' (1864). The other parent is not known, but it shows more of *R. gigantea* than its parent and it may have been recrossed with Parks's Yellow Tea. The popular 'Gloire de Dijon' was classed as a Noisette by nineteenth-century writers on the subject, but appears to be rather a climbing Tea. It is known that one parent was a Bourbon Rose, 'Souvenir de la Malmaison'. This is a very vigorous rose with large pale pink flowers and may well have been the principal influence in the climbing 'Gloire de Dijon', the other parent being a yellow Tea Rose.

Just as the first remontant hybrid came from outside Europe, so did the second, the Bourbon Rose. This appeared as the second generation of a cross between the Autumn Damask and Parson's Pink China and originated in the island of Bourbon in the Seychelles. It differs from the Noisette Rose in having *R. gallica* in its parentage in addition to the other three species. The first Bourbon Rose was distributed in France in 1823 and reached England two years later. Although the Bourbons were cultivated on their own account they were also crossed with the early tea-scented Roses, Hume's Blush and Parks's Yellow to produce the Tea Rose. Most Bourbon Roses are triploids with twenty-one chromosomes; however they appear to cross quite happily not only with each other, but also with the *chinensis* X *gigantea* Roses. 'Gloire de Dijon' is for some reason a tetraploid with 28 chromosomes. The first genuine Tea Rose was 'Smith's Yellow' bred by crossing the blush Noisette with Parks's Yellow. This was in turn backcrossed with Parks's Yellow to produce that beautiful lemon yellow Rose 'Devoniensis' (1838). In 1839 Parks's Yellow was crossed with 'Jaune Desprez' and one seedling was known as 'Safrane' with outer petals which were deep pink, while the inner ones were bright yellow. The pink Tea Roses seem to have arisen by crossing Hume's Blush with the Bourbons, while the Noisettes were also crossed with the Bourbons to give purple and crimson shades. These when crossed with Hume's Blush enriched the colour of the Tea Roses further. The only disadvantage from some gardeners' point of view was that the large amount of *R. gigantea* blood in the Tea Roses made them tender in northern climes. In England they had to be grown in glasshouses or sheltered situations and most of them were bred in central and southern France.

In the early nineteenth century the various China Roses were crossed with all the other cultivated roses and were known as Hybrid Chinas. Most of those that have survived prove to be triploids, but there were four which were tetraploid and became the parents of the Roses known as Hybrid Perpetuals. These four hybrids were a *chinensis* X *gallica* cross known as 'Malton' and three Bourbon X *gallica* crosses 'Brennus', 'Athalin' and 'General Allard'. These four Roses all date from about 1830. The Hybrid Perpetuals were the result of crosses between these

Hybrid Chinas and the Bourbons and Noisettes. Indeed the first Hybrid Perpetual, the 'Rose du Roi', seems to have been raised from the Portland Rose. The majority of the Hybrid Perpetuals came from the three Bourbon X *gallica* hybrids. These were either selfed or crossed with the Bourbons and Noisettes. The 'Rose du Roi' was raised in 1816, but in 1837, Laffray, a nurseryman at Auteuil, near Paris, raised 'Princesse Hélène', with much larger flowers than those in which the Portland Rose had been concerned. Although the results of the selfed Hybrid Chinas were all sent out as Hybrid Perpetuals, some of them were, naturally, not perpetual flowering. Others, such as 'Jules Margottin' (1852) a selfed seedling of the Bourbon X *gallica* 'Brennus', were remontant.

It was natural to cross the Hybrid Perpetuals with the Tea Roses and the resultant roses are known as Hybrid Teas and these remain the most popular Roses to this day, although, as we shall see, the modern Hybrid Teas are somewhat different from the originals. The first Hybrid Tea was a cross between the Hybrid Perpetual 'Victor Verdier' and the Tea Rose 'Mme Bravy' and was known as 'La France'. This remarkably lovely Rose is still in cultivation a hundred years after it was introduced. The Hybrid Tea combined the graceful high-centred shape of the Tea Roses with the greater vigour and hardiness of the Hybrid Perpetuals. They contain in various proportions strains of *RR. gallica, moschata, chinensis* and *gigantea* and it is not impossible that *R. phoenicea* and *R. canina* have also got in through Summer Damask and *alba* Roses, so that they are quite a confused group with abundant possibilities of variation. The first Hybrid Teas were all raised by crossing Tea Roses with Hybrid Perpetuals, but later the Hybrid Teas were crossed with each other and also back-crossed to Hybrid Perpetuals, thereby giving rise to such floriferous red roses as 'Richmond'. All the Hybrid Teas are tetraploid with twenty-eight chromosomes. The Teas were diploids with fourteen chromosomes while the Hybrid Perpetuals are tetraploids like the Hybrid Teas. Why the Hybrid Teas are not triploids is puzzling, or would be with any species less liable to cytologic eccentricity.

In 1860 seeds of the climbing *R. multiflora* were received at Lyon and the plants appear to have been accidentally fertilized with the dwarf Pink China Rose, known as *R. lawranceana* or the 'Fairy Rose'. In the F2 generation two dwarf pompom Roses segregated out. These were the first Polyantha Roses. Subsequently these Polyanthas were crossed with Hybrid Teas to give the Floribunda roses. Although there had been crosses between Polyanthas and Hybrid Teas in the 1880s ('Cecile Brunner' and 'Perle d'Or') the first real Floribundas were created by the Danish breeder Poulsen. His first seedling 'Else Poulsen' appeared in 1924.

Before the first Floribunda Roses were bred the Hybrid Teas had been transformed and this transformation was entirely due to the activities of one breeder, Pernet-Ducher of Lyon. He wished to enrich the colouring of the existing roses by breeding into the strain *Rosa foetida.* This exists in two clearly defined colour forms. *R. foetida* itself, the Austrian Briar, the only rich yellow Rose known at the time; and another form, the Austrian Copper *R. foetida bicolor,* which is a mixture of yellow and scarlet. As we have noted, the yellow form had been bred into the Burnet Roses quite early in the nineteenth century, but *R. foetida* was nearly sterile, and it was therefore very difficult to obtain any good pollen. Pernet-Ducher was using the double form of *R. foetida,* which did not make his task any easier as there was less pollen available and of this probably only five per cent was fertile. He attempted to cross various Hybrid Perpetuals with *R. foetida* between 1883 and 1888 and in his last year obtained some fertile seed on a large red Hybrid Perpetual 'Antoine Ducher'. From the few seeds obtained two plants resulted. One was completely sterile and known as 'Rhodophile Gravereaux'; the other plant was never named but is known as the Lyon Rose. In shape the flower was like a Hybrid Perpetual but it had the bicoloured flowers of *R. foetida bicolor,* an orange-red on the inside and yellow on the outside. The Lyon Rose was not distributed but used as a pollen parent on various Hybrid Teas, in order to bring the remontant habit back in, and the first Rose to be distributed was 'Soleil d'Or'.

Pernet-Ducher's aim was to bring into the Hybrid Teas a richer yellow than could be obtained from *R. gigantea,* but he was taken by surprise by the various bicoloured effects that the Austrian Copper form introduced. He was successful in introducing a good yellow colour, but there were various undesirable side-effects as a result of introducing *R. foetida* into the hybrids. *R. foetida* is slightly tender in this country and this tenderness was imparted to the first Pernet Roses, making them inclined to winter die-back and general lack of vigour. In addition *R. foetida* is very susceptible to the fungus disease known as Black Spot, which is characterized first by black spots on the leaves and then by their falling prematurely. Before Pernet's success this had not been a serious complaint, whereas now it is practically universal except in towns, where the atmosphere is sufficiently polluted to prevent the spores spreading. The first Pernet Roses were also a bad shape, compared to the existing Hybrid Teas and *R. foetida* brought an excessive number of prickles into the garden roses. The Austrian Briar is also, apparently, responsible for the shining foliage of the latest Hybrid Teas, although *R. foetida* itself has rather dull leaves. Another disadvantage was the fact that fertility was very low and the raising of seedlings was very difficult in the early stages. This disadvantage has now been overcome. Since Pernet-Ducher's day several other yellow Rose species have been introduced from Asia and it might well be worth some breeder's while to breed such Roses as *R. hugonis* and *R. xanthina* into pre-Pernet Roses. These could result in equally good yellows and plants that were less susceptible to Black Spot. On the other hand this pest has now become so widespread that it may be too late.

The species that have gone to make the modern Floribunda rose are therefore *Rr. gallica, moschata, chinensis, gigantea, foetida,* and *multiflora*; and probably *phoenicea* and *canina* can be added to these, making eight species in all. There are not many garden plants with so involved an ancestry.

It will be recollected that the remontant habit of *R. chinensis* was due to a recessive gene that caused the climbing shoots to become dwarf flowering stems. Occasionally this mutates back and we get climbing forms of Hybrid Teas.* These are usually once-flowering, but they may produce a few extra flowers later in the season. In the meantime breeders had also been working on producing coloured climbing roses. Nothing very outstanding was produced either from *Rosa arvensis* or from *R. sempervirens*, but *R. multiflora* and *R. wichuraiana* (*R. luciae*) have been extensively used. It is from these species, crossed with various Teas, Hybrid Perpetuals and Hybrid Teas that the various Ramblers have evolved. Hybrids with *R. wichuraiana* are characterized by extremely glossy foliage. At one time there were other climbing roses, that have now become extinct or extremely rare. The Prairie Roses were hybrids with the late flowering North American *R. setigera*, while the Boursault Roses, which are thornless, have long been thought to have been hybrids of *R. pendulina*, the Alpine Rose. The fact that *R. pendulina* is a tetraploid makes this rather improbable and Dr Hurst suggests that some American species is responsible. The Boursaults lack fragrance and this also suggests that the Alpine Rose, which is extremely fragrant, was not a parent. The very fragrant thornless 'Zephirine Drouhin' is presumably a cross between a Noisette and a Boursault. The various *multiflora* and *wichuraiana* ramblers have been crossed with both Hybrid Teas and Floribundas to give rise to a race of so-called Pillar Roses or Climbing Floribundas. These are much more controlled in growth than the Ramblers, rarely reaching a height of over eight feet. On the other hand they have remontant flowering, which is an advantage. Among some of these is a species we have not met before, *R. rugosa*, the Japanese Briar. This is an attractive and thrifty species with sweetly-scented blooms and handsome heps; it hybridizes readily with other species but the resultant hybrids have all proved sterile. This must be due to some incompatibility in the chromosomes as they have the usual diploid count of 14. In 1919 a hybrid between *R. rugosa* and *R. wichuraiana* had been raised and named 'Max Graf'.

* This climbing habit is often found only in the outer parts of the plant. Root cuttings of climbing HT's often produce the original bush form.

Sometime in the late 1930s or early 1940s a tetraploid form occurred and the seedlings from this have been named *R.* X *Kordesii* after its raiser, the great Rose breeder Wilhelm Kordes. This has been used in such Pillar Roses as 'Leverkusen' and 'Parkdirektor Riggers'.

This would seem a suitable place to mention the various shrub Roses raised by Herr Kordes in recent years. Although they number many different hybrids his products can be especially signalled by their re-use of two old species, the Sweet Briar and the Burnet Rose (*RR. rubiginosa* and *spinosissima*).* These are mostly single flowered and single flowering shrubs of great beauty. The *spinosissima* hybrids result from crosses with Hybrid Teas and have produced such exquisite flowers as 'Frühlingsgold' and the other 'Frühling' series. The Hybrid *rubiginosas* come from various roses crossed with 'Magnifica'. This was the result of selfing one of the Penzance roses 'Lucy Ashton', which was a cross between *R. rubiginosa* and an unknown Hybrid Perpetual. There is thus rather less of *rubiginosa* blood in these roses than there is of *spinosissima* in the Burnet hybrids. Some of the resultant roses are remontant, but most are once-flowering but of great vigour and brilliance. Most of the remontant plants are classed as floribundas, although they differ from most of these by having 'Magnifica' in their ancestry.

The Hybrid Musk Roses (although far removed from *R. moschata*) are the outcome of the efforts of two breeders separated in time and nationality. We start with a German breeder, Peter Lambert. He started work with the cultivar 'Aglaia'. This is a climbing Rose caused by crossing *R. multiflora* with the Noisette 'Reve d'Or'. This was reputedly crossed with a Hybrid Tea 'Mrs R. G. Sarman-Crawford', but doubts have been thrown on this and it is thought that the resultant Rose, christened 'Trier' was in reality a selfed seedling of 'Aglaia'. Lambert crossed 'Trier' with various Hybrid Teas and produced a series of roses, known in their day as Lambertiana

* Though Herr Kordes states that he used *R. spinosissima*, it is not in fact clear whether he used that species or *R. pimpinellifolia*.

Roses, but now little seen. The Reverend Joseph Pemberton in turn crossed these Lambertiana Roses with various Hybrid Teas and Hybrid Perpetuals and produced a number of roses, generally with delicious fragrance, in a variety of styles from erect shrubs to climbers needing support. Although they could be described as remontant, they normally have two main floral displays, at midsummer and in the autumn with a few blooms in the interval. The flowers are smaller than Hybrid Teas but larger than most *multiflora* hybrids. The shrub forms make large shrubs in time and they fill a place that no other hybrids have so far accomplished.

In certain books devoted to Roses, written in the early and mid-nineteenth century, there are references to Microphylla Roses. These are based, not on *R. roxburghii* (*microphylla*), but on a plant figured on Plate 919 of the *Botanical Register*. This shows a double pink rose, with leaves similar to those of *R. willmottiae* and very long bracts with prickles at the base. It would seem, therefore, to show some similarity to *R. roxburghii*, which has a prickly calyx. Many contemporary botanists thought that it had some connection with *R. bracteata*, but Lindley, who provides the text accompanying the illustration, thinks that it is a form of *R. chinensis* (which he calls *indica*). The plant seems to have been obtained from the Calcutta Botanic Garden and flowered in this country in 1822. It looks as though it might be a hybrid between *R. chinensis* and *R. roxburghii*. A new Microphylla Rose was bred as recently as 1868, but their popularity was already on the wane. Shirley Hibberd in the *Amateur's Rose Book* suggests that they are slightly tender, but this may be because he thought they were hybrids of the tender *R. bracteata*. Unless some plants still survive in cottage gardens, this race would appear to have vanished. According to Graham Thomas (*Shrub Roses of To-day*, London, 1962) the original importation is *R. roxburghii*, 'an ancient cultivated form of the "Chestnut Rose" . . . presumably of Chinese origin'. The wild form of *roxburghii* (var. *normalis*) is said only to have been introduced in 1908. However both Sweet and Loudon mention a *R. rox-*

*burghii* in their 1830 publications. Loudon says that it was white-flowered and introduced in 1821, before '*microphylla*', so the matter requires further investigation. *R. roxburghii* var. *roxburghii* is still in commerce.

At the beginning of the century collectors such as Wilson and Forrest had introduced to cultivation a number of species from China. Many of these, such as *Rosa moyesii* and *R. hugonis*, are delightful shrubs in themselves and it might be expected that they would have been used in the breeding of fresh hybrids. This has been done to only a limited extent, but mention must be made of the cultivar 'Nevada' a cross between a Hybrid Tea 'La Giralda' and some form of *R. moyesii*. *R. moyesii* itself is a hexaploid ($2n=42$) but there are tetraploid forms (*R. moyesii fargesii* and *R. holodonta*) and either of these could be crossed successfully with a Hybrid Tea. In any case *R. moyesii* seems capable of improbable crosses. 'Eos' is a cross between *R. moyesii* and the pentaploid 'Magnifica' and this cross is fertile enough to cross with 'Kordesii' giving the cultivar 'Bengt M. Schalin'. It must be accepted that Roses have a peculiar cytology and that crosses that would be impossible with most plants can be performed with ease by this genus. Why a selfed seedling of a hexaploid 'Lucy Ashton' should give rise to a pentaploid in 'Magnifica' is baffling enough in all conscience, but that this pentaploid should be a good parent is even more improbable. Very often in reading the pedigrees of Roses you read that such a cultivar is reputed to be a cross between A and B, but the chromosome count makes this unlikely. With any other genus one would probably read 'makes this impossible'.

The breeding of a blue Rose is always held up as being the rose breeder's ideal. On the face of it it does not seem very probable. Blue is not found at all, so far as I know, among the *Rosaceae*. On the other hand the scarlet pigment pelargonidin has been bred into roses and sometimes this seems to be recessive to blue, as in *Anagallis*, so that it is just remotely possible that such a flower could be bred. There seems no reason to suppose that it would be particularly attractive, any more than those other desiderata of the hybridist the red Daffodil and the yellow Sweet Pea would be. Indeed the red Daffodil has nearly been accomplished and seems rather unpleasing. There are some Roses that are more or less violet in colour and these have not been received with much favour.

The latest novel trend in Rose breeding has been the creation of shrubs that will have the grace of the species combined with the remontant habit of the Chinese Roses. One such is 'Erfurt' which is the result of crossing 'Eva' – a hybrid between a Hybrid Musk, a Polyantha and a Hybrid Tea – with the Hybrid Tea 'Reveil Dijonnais' and which bears single pink flowers shading to yellow in the centre. Another of these perpetual shrub Roses is 'Golden Wings', a hybrid between 'Ormiston Roy', a *spinosissima* hybrid, and 'Soeur Therese', a hybrid of a Hybrid Perpetual with a yellow Hybrid Tea. This gives good, light yellow, open flowers that are produced continuously.

It is difficult to foresee what the next development will be. Obviously more Hybrid Teas and Floribundas will be produced to meet the insatiable demand and the main hope for further developments in this field would seem to be the introduction of new species. 'Cerise Bouquet' has crossed the Hybrid Tea 'Crimson Glory' with *Rosa multibracteata* and one can conceive that some of the small-flowered species such as *Rr. farreri persetosa* and *R. willmottiae* could be used to breed a race of small-flowered Roses. The possibilities still seem infinite.

CHAPTER VI

# Diianthus

Both the Pink and the Carnation have been favourite cultivated plants for so long that their early development took place before any botanical writings.

### ❀ *Carnation*

The Carnation is derived from *Dianthus caryophyllus*, a sub-shrubby plant with flowers in clusters, up to an inch across and with notched petals. Owing to its long history of cultivation, its country of origin is not known with any certainty. According to Pliny it was found in Spain in the reign of Augustus and was used to flavour wine. The epithet *caryophyllus* refers to its clove-like perfume.

A good many wild forms of *D. caryophyllus* are found in the Atlas Mountains, a fact which lends colour to the legends of its introduction from Spain in Roman times and its original exploitation by the Moors. It is said to have been introduced into England by the Normans and it is certainly still naturalized on the walls of such Norman castles as Rochester, besides being present on the walls of William the Conqueror's castle at Falaise. There is also a tradition that it was first developed by the Moors and introduced to Europe from Tunis in the thirteenth century. These were probably the first of the doubles that became so popular among gardeners. The type of *D. caryophyllus* is a deep pink in colour and it is possible that other species were bred in from time to time, producing the many kinds of Border Carnations. Certainly the yellow-flowered plant that Master Nicholas Lete, that 'worshipful merchant of London' procured from Poland and gave to Gerard suggests that the yellow Dianthus – *D. knappii*, a native of

Hungary – had contributed its colour. In the sixteenth century the boundaries of Central Europe were rather elastically drawn and Poland would probably include the homeland of *D. knappii*. In that century double Dianthus, looking more like Pinks than Carnations, were also frequently depicted on Persian tiles and so it is possible that here again the Turks and Persians were the originators of garden forms of these plants. A curious feature of the early Carnations is the predominance of striped and mottled flowers. This striping is not the result of a virus as in the case of the Tulips and it is not recorded in wild plants. Presumably it is the result of selection, but how the break first occurred is lost in the past. These Flakes, Bizarres and Picotees appear to be just as vigorous as the self-coloured forms of the Border Carnations and long-established clones show no sign of weakness.

The doubleness of the Carnation is caused by the stamens becoming petaloid; the stigma is intact and the flower can be fertilized. Moreover some of the doubles also have some stamens and can be used as pollen parents as well. Those flowers that are completely stamenless tend to burst their calices and so are not much regarded as commercial plants, but crossed with singles or semi-doubles these plants are often excellent parents.

An early development of the Carnation were the Remontant Carnations, the first of which appears to have been raised at Ollioules in southern France around 1750. They were known as Oeillets Mayonnais or Oeillets de Mahon, and flowered from June until September. In 1830 M. Dalmais, a gardener in the nursery of Lacême of Lyon, crossed an Oeillet de Mahon with a Border Carnation called 'Bisbon', and the resultant flower was the first of the true perpetual-flowering plants. This plant was crossed with other Border Carnations and a number of different colours were raised. We may well ask ourselves what caused the remontant character in the Oeillets de Mahon. The late Montague Allwood, who spent his whole life raising different forms of Dianthus hybrids, was of the opinion that the perpetual-flowering Carnation was a hybrid of *D. caryophyllus* with *D. chinensis*. He also thought that other species might be involved, but did not specify them. If he is correct it is fascinating to realize that China was responsible not only for the remontant Rose, but also for the perpetual-flowering Carnation. *D. chinensis* is an annual or biennial plant that is very floriferous and there seems nothing improbable in thinking that if *D. caryophyllus* were pollinated with *D. chinensis*, the result might be a perpetual-flowering perennial. The opposite cross with *D. caryophyllus* pollinating *D. chinensis* has given rise to the annual Marguerite Carnations, which show a similar remontant quality but have broader and greener leaves than *D. caryophyllus* itself has. Apparently Mr Allwood never made the cross of a Border Carnation with *D. chinensis* to prove his point and no one else seems to have done it subsequently. Like Dr Hurst's theoretical origins of the Damask roses, it sounds very plausible but still requires practical demonstration. Although *D. chinensis* behaves like an annual, nineteenth-century gardeners stated that if kept under glass and prevented from setting seed, it would perennate. If this is so, it would explain the perennial quality of the Perpetual Carnation. Miller in his *Gardener's Dictionary* refers to a plant he calls the 'Old Man's Head Pink' which sounds a bit like the 'Oeillet de Mahon'. He says that it does not flower until July 'coming at the same season with the Carnation, to which it is more allied than the Pink' and that 'this sort will continue flowering till the frost in autumn puts a stop to it'. Unfortunately we do not know when the 'Old Man's Head' was introduced. Since *D. chinensis* reached Europe in 1705 and England in 1722, there was nothing to stop this cross having taken place in England as well as in France. On the other hand it would seem that the 'Old Man's Head' was hardy in England, which is more than can be said for the Perpetual Carnation.

After its original appearance in Lyon in 1830, breeding of the Perpetual Carnation continued almost exclusively in France for many years; the greatest of the early breeders was Alphonse Alegatière whose best work was done in the 1860s and 70s. At the same time American breeders had imported seeds and plants from France and were breeding on their own account.

It seems odd nowadays to read that it was not until 1890 that florists started disbudding the Carnation, so that each stem bore only a single flower, thereby causing an increase in size and symmetry. In the history of the Carnation, it is the early American cultivars whose names have survived, although 'L'Alegatière' is still remembered as one of the best French cultivars. Of these early cultivars the most important was 'Mrs Thomas W. Lawson' raised by Peter Fisher from a hybrid between 'Daybreak' and 'Van Leeuwen'. This appeared in 1895 and has been one of the most frequently used parents of later cultivars. It was a cerise-pink self and had a more marked remontant habit than any previous cultivar. Mention should also be made of an earlier cultivar 'Victor Emmanuel' raised about 1860 by Donati, which was a yellow-ground Picotee and probably the parent of subsequent flowers of this colour.* Another cultivar of Peter Fisher's was the large pale pink 'Enchantress' but this, though a handsome flower in itself, proved sterile both as male and female.

M. Allwood in his classic *Carnations, Pinks and All Dianthus*† points out that the British breeders were behind the times, because they originally tried crossing the Perpetual Carnation with the best Border Carnations, thereby producing single flowering plants. However the cultivar 'Miss Joliffe' a pink-flowered plant was used extensively in America as a parent and since it was the pollen parent of 'Daybreak' it can be regarded as a grandparent of 'Mrs T. W. Lawson'. Another important early British cultivar was 'Winter Cheer' raised by the James Veitch nursery. This was, in spite of its name, not particularly floriferous in the winter months; Mr Allwood says that it must have been a Perpetual X Border cross, but was a compact plant with a good scarlet flower. This was crossed in 1904 with 'Mrs T. W. Lawson' to give 'Britannia', still regarded as one of the best of the early Carnations. Apart from its good scarlet flowers, it was the first scarlet that did not lose its colour during the winter and it was markedly more floriferous than most of the

* Although early lists of cultivars include many yellows.

† 3rd edn., 1947.

previously raised cultivars. Moreover there are few of the darker-flowered Carnation cultivars even today that do not have 'Britannia' somewhere in their parentage.

In 1857 a certain Monsieur Lasie was raising seeds of Perpetual Carnations and one of his seedlings had flowers that were markedly larger and more handsome than in any previous cultivar. Owing to its likeness to a rose that was popular at the time he gave it the same name 'Souvenir de la Malmaison'. Unfortunately it would not set seed and produced only a very little pollen, which also did not seem viable. In 1876 three budsports were produced and propagated; one of these, 'Princess of Wales', became a flower of fashion at the end of the century. Although the Malmaison Carnations were so spectacular they were not easy to grow and were sparing of their flowers. In 1906 Mr H. Burnett of Guernsey produced 'Marmion' a white-edged red Perpetual Malmaison, but this was unfortunately completely sterile. Martin Smith used to cross the original Malmaisons with Border Carnations and got some very striking results. His triumph was 'Mrs Martin Smith' which, according to Allwood, grew five foot high and had flowers eight inches in diameter. Unfortunately these spectacular blooms were produced only very sparingly. As late as 1951 Mr Allwood was successful in combining the Malmaison with the Perpetual Carnations giving rise to a race that he called Royal Carnations. Some idea of the difficulty can be gathered from the fact that he fertilized 300 flowers from which he got only 32 seeds. Fortunately plants from these seeds were fertile and it was possible to develop the strain further; this was done most notably in a series of crosses with a Picotee 'Royal Fancy', which gave rise to a number of exceptionally large perpetual-flowering Picotees. A further early cultivar which should be noted was 'White Perfection' raised in the U.S.A. by F. Dorner at the end of the last century. A cross between this and 'Britannia' gave the cv. 'Wivelsfield White' which remained popular for fifty years and is still in cultivation.

## ✿ *Pink*

The Pink is derived from *Dianthus plumarius,* a native of the calcareous mountains that stretch from the Italian Alps into Austria, Hungary and the Tatra. The plant usually bears only a single flower on each stem and the wild forms are white or deep pink, sometimes with a central blotch (var. *zonatus*). It was from the form with this central blotch that in 1772 James Major bred his 'Duchess of Lancaster', the first 'laced' Pink.* In Laced Pinks the edges of the petals are tipped with the same colour as the central blotch and the ideal of the breeders was an almost black centre and broad lacing.

It has been suggested that the original Pinks were probably hybrids of *D. plumarius* with other species, notably *D. gratianopolitanus,* the Cheddar Pink. It seems fairly likely that the Border Carnation also made some contribution to the Laced Pinks. C. H. Herbert collected a lot of the old Paisley Laced Pinks (the Laced Pink was grown extensively in the mid-nineteenth century by the Paisley weavers) and bred from them with results similar to those obtained by Montague Allwood when he crossed Pinks and Border Carnations. The race known as 'Allwoodii' was created by the same breeder after nine years' intensive work on crossing Pinks with Perpetual Carnations, that is *D.* (*plumarius* X *gratianopolitanus*) X *D.* (*caryophyllus* X *chinensis*). Allwood's Show Pinks were the result of crossing Herbert's Laced Pinks with 'Allwoodii'.

The last of Allwood's Dianthus hybrid races was the 'Allwoodii alpinus' range, bred by crossing 'Allwoodii' with various dwarf species, notably *Dd. alpinus, gratianopolitanus, arenarius* and *superbus.* It is a race of dwarf, caespitose pinks and was the result of nine or ten generations of crosses. The plants flower over a long period, but lack much variety in colour. Most of them are pink or purple.

C. H. Fielder raised a race of Pinks that he called Lancing Pinks, compact plants with a long flowering season. These were crosses of 'Allwoodii' with a tufted plant given as *D.* 'Winteri'.

* Not 'Lady Stoverdale' of 1774 as usually stated. See *Floricultural Cabinet,* 1841.

*Dianthus* 'Winteri' is described in *The New Flora and Silva** as a race of single remontant flowering pinks raised by a Mr Winter, a dweller in East Anglia. The parents seem to have come from Sydney Morris of Earlham Hall, who is remembered for the Earlham strain of Montbretias. Mr Winter received two seedlings in 1922 and crossed them. Another Earlham Hall seedling with flowers up to 2½ inches in diameter was then bred into this strain and further crosses and selections were made, until the cvs. were introduced to commerce in 1932. Although all the plants were single, they were valued because of their compact habit and long flowering season, but they were soon superseded by double cvs. and never became as popular as the original describers had hoped. Their parentage is not sufficiently well-known to explain their habit of continual flowering. It seems possible that it might have been *plumarius* X *chinensis*: several of the cvs. are described as having deeply laciniated petals and this certainly suggests *D. chinensis*. On the other hand there are other laciniated petalled species and it might be that *D. superbus* entered into them. The secret seems to be buried with Sydney Morris.

### ❀ *Sweet William*

*Dianthus barbatus*, Sweet William, a native of southern Europe, has a long history of cultivation behind it, but has changed comparatively little during its centuries in the garden. There are doubles and there is also a salmon pink shade, which is uncommon in the wild. The crimson and the white with a purple centre are those most commonly found in the wild. However the plant has been in cultivation so long that it is not easy to decide whether a plant is truly wild or a garden escape.

In 1717 Thomas Fairchild produced the so-called Mule Pink by crossing *D. barbatus* with a Carnation. This hybrid had double red flowers but was sterile. 'Allwoodii' crossed with *D. barbatus* gave rise to the hybrid called 'Sweet Wivelsfield', while this crossed with the deeply-fringed *D. superbus* var. *speciosus*

* By J. L. Gibson, Vol. 3, p. 44 (London, 1930).

gave the race known as 'Rainbow Loveliness'. All these hybrids are due to the late Montague Allwood as also was 'Delight', raised by crossing 'Sweet Wivelsfield' with the dwarf *D. neglectus* var. *roysii.*

The haploid number for *Dianthus* is 15 and most species are diploid with $2n=30$. However the cultivated Pink and Carnation are hexaploids with $2n=90$. 'Allwoodii' have the same count. *D. chinensis* exists in diploid, tetraploid and hexaploid forms. *D. gratianopolitanus* in tetraploid and hexaploid forms. The Laced Pink is a hexaploid. *D. knappii* is a diploid, which explains why it is not easy to breed into these polyploid hybrids.

The Border Carnation was raised to a state of perfection at a very early date and *D. chinensis* had an even longer history of cultivation behind it, so, assuming its putative parentage is correct, the Perpetual Carnation is the result of the crossing of plants that had been garden favourites for a long time. From the decorative point of view the hybrid seemed to slow down the pace of development and it is only within living memory that the Perpetual Carnation has had flowers that were as decorative as the old Border Carnations.

It is never easy to foresee future developments, but a Perpetual flowering laced Carnation has yet to be developed. Mr Allwood had a nearly blue Dianthus sent him from China in 1934; apparently the plant had been given no name by 1954. It was similar in appearance to Sweet William. If this plant has survived, it evidently has great possibilities. Allwood was working on climbing Dianthus and climbing Carnations, although I cannot see that these would have anything other than curiosity value. He had also been trying to raise a bush Dianthus and this would seem to have possibilities. The so-called Tree Dianthus, *D. arboreus,* is a small shrubby plant about 18 inches high and coming from the Aegean is tender, but could possibly be used as a parent. There might be advantages in obtaining wild plants of *D. caryophyllus* and *D. plumarius,* which might introduce some fresh genes into the hybrid strains with satisfactory results. In any case this seems a genus that still has great potentialities.

## CHAPTER VII

# The Iris

It is a relief to turn from the conjectures of the Dianthus to the well-documented history of the various Iris hybrids. The genus *Iris* manifests itself in many forms and nearly all have been extensively developed by breeders. However, what most people mean when they speak of Irises is the race known as Tall Bearded and we may well consider these first.

There are a large number of apparent species of bearded Iris in Southern Europe and the Middle East, but cytology has demonstrated that not all of them are genuine species. These include Linnaeus's *Iris germanica,* thought by many to be the ancestor of all our rhizomatous garden irises. In point of fact *I. germanica* is a sterile clone with a chromosome count of 44. The same count and the same sterility occur in the source of orris root, *I. florentina,* and in the white Iris, *I. albicans,* that is found around Moslem cemeteries. The normal diploid count of the Tall Bearded Irises is 24 and these 'species' are regarded as hybrids between a dwarf Iris such as *I. chamaeiris* with a diploid count of 40 and a tetraploid tall bearded ($2n=48$), of which many species exist. Most of these tetraploid Irises come from the Middle East and it may be assumed that the hybrids must have originated in the eastern Mediterranean. The two species that formed the ancestors of our garden Tall Bearded Irises are both diploids; they are *Iris pallida* and *Iris variegata,* both most commonly found around the Adriatic and hybridizing in the wild as well as in cultivation. Some of these natural hybrids were given specific names by nineteenth-century botanists, such as *II. amoena, plicata, sambucina* and *squalens* and these were reproduced artificially by such breeders as Foster, Bliss and Dykes.

Although there seems no reason to doubt that *II. pallida* and *variegata* are the main parents of the Tall Bearded Iris, there are some allied species *II. cengialtii, imbricata* and *illyrica,* which may also have played their part. *Iris pallida* is a tall species with fragrant flowers in varying shades of lavender, while *I. variegata* has yellow standards and brownish falls.

As far as is known the selection of hybrid Iris seedlings was started about 1800 by a German, E. von Berg of Neuenkirchen, and a Frenchman, de Bure of Paris. Von Berg was apparently an amateur and though he named and described his seedlings, he did not distribute them; on the other hand de Bure was a commercial grower and distributed the plants. The most famous, a flower with a pale ground stippled over with a darker colour of the type now known as 'plicata', he named after himself 'buriensis'. It was distributed about 1822 and survived in cultivation for a century. De Bure was followed by another French nurseryman, a M. Jacques, one of whose seedlings 'aurea' remained the best yellow self until 'W. R. Dykes' was introduced in 1929.

Jacques's work was continued by the nurseryman Lemon, who in 1840 was offering a hundred named varieties. Many of his cultivars, particularly 'Jacquesiana' and 'Mme Chereau' remained for a long time in cultivation and may even now be seen occasionally. Lemon continued to offer new cvs. for fifteen years and it is principally owing to him that the Tall Bearded Iris became a popular garden flower. Naturally other nurserymen followed suit and we find new Irises being offered by such well-known nurserymen as Van Houtte, Victor and Eugene Verdier, John Salter and later Peter Barr, Amos Perry and George Reuthe.

The situation was transformed at the end of the century. Sir Michael Foster, a professor of physiology at Cambridge, was a keen amateur breeder. He felt that there was little chance of improvement with the existing strains and asked friends of his who travelled in the Middle East to send him any tall bearded Irises that they found on their travels.

Since most of his correspondents were missionaries and not

botanists, these plants arrived without any labels and were named from their provenance. Thus one from Troy was named *I. trojana*, one from Amasia has been given the improbable name of *I. amas* (or *I. germanica* var. *amas*) while Iraq and Cyprus provided *I. mesopotamica* and *I. cypriana*. These were all rather tall Irises of the 'germanica' type in flower colour, varying shades of lavender and purple but they were all tetraploids with a chromosome count of 48. One of the curiosities of the pogon (bearded) Irises is the way that cell reduction frequently seems to be in abeyance. One would expect a hybrid between a diploid and a tetraploid to be a sterile triploid, but this was not invariably the case. In time breeders such as Foster himself, Dykes, George Yeld and Amos Perry bred a race of tetraploid Tall Bearded Irises which has now to all intents and purposes completely supplanted the original diploid race. As might be expected from its provenance *I. mesopotamica* is rather tender for this country and the northern U.S.A. and it has handed on its tenderness to its offspring, which include some of the most valuable parents.

The true specific validity of these tetraploids is somewhat in doubt and they may be the results of early Middle East gardening. However, it is more convenient to treat them as species.

Another Iris species which was used is a white Iris from India, *I. kashmiriana*. This is a rather curious species that appears to exist not only in diploid and tetraploid forms but also as a sterile clone with a count of 44 chromosomes. It is of great importance in breeding as its white colour is dominant, whereas most whites are recessive.

Another odd thing about the Tall Bearded Iris is their ability to acquire or shed a few chromosomes without their fertility being impaired, as can be demonstrated by the pedigree of 'Purissima' one of the most valued parents and still one of the best white Irises yet raised. On the female side we start with a cross 'Caterina' between *I. cypriana* (48) and *I. pallida* (24). 'Caterina' had 48 chromosomes and was crossed with *I. mesopotamica*, also with 48 chromosomes and the offspring

'Argentina' has 50 chromosomes. On the male side we have a diploid cv. 'Juniata' (24) crossed with *I. mesopotamica* (48) to give 'Conquistador' with 49 chromosomes. 'Argentina' crossed with 'Conquistador' gave 'Purissima' with 47 chromosomes.

With the chromosomes behaving so obligingly it is not as surprising as it would be in most other genera to learn that a race of Iris with 20 chromosomes has been successfully bred into the Tall Bearded. This is the spectacular group known as Oncocyclus, restricted to a small area in the Middle East from southern U.S.S.R. to Iran and of great difficulty in cultivation. Here again the pioneer work was done by Foster whose hybrids included 'Parvar' (*paradox* X *variegata*) and 'Ib-Pall' (*iberica* X *pallida*). These date from the end of the nineteenth century. Of greater importance in breeding has been van Tubergen's cross of 'Ib-Mac' (*iberica* X *amas macrantha*. This last plant is sometimes referred to in Iris pedigrees as 'Macrantha' treated as a cv. name). Although, as we shall see, this plant has been a valuable parent, it is not to be met with in catalogues. The first commercially introduced hybrid between an Oncocyclus and a Tall Bearded bloomed for the first time in 1923. This was raised by the famous breeders William Mohr and Sydney Mitchell and was named after the former by the latter. 'William Mohr' was a cross between the diploid Tall Bearded 'Parisiana' and *Iris gatesii*, a spectacular plant from southern Turkey. There seems to be some slight confusion as to which was the pollen and which the seed parent. As would be expected 'William Mohr' has 22 chromosomes and is a bad setter of seeds. Moreover it very rarely produces any pollen at all. Nevertheless there were bred a certain number of descendants, all with the word Mohr in their cv. names.

A more fruitful source of imparting Oncocyclus blood into the Tall Bearded came in 1940 when Frank Reinelt flowered 'Capitola'. The parentage of this was 'William Mohr' X 'Ib-Mac'. 'Ib-Mac' has 44 chromosomes and 'William Mohr' has 22 but 'Capitola' has 43 and not 33 as would have been expected. 'Capitola' has proved an extremely potent pollen parent and the number of Oncobreds, as hybrids between the Tall Bearded and

Oncocyclus are termed, has risen to a marked degree since its introduction. Oncobreds are entirely an American strain at the moment and they cannot be relied upon to flower freely in this country. The Oncocyclus strain needs a hot summer to ensure that the rhizomes ripen properly. As far as I can find out no one has tried crossing the 44-chromosome 'Ib-Mac' with the 44-chromosome Irises known as *florentina, albicans* and *germanica.* Perhaps the chromosomes are so disparate that they would not recombine, but it seems a possible field of action. American breeders are now trying to introduce other Oncocyclus Iris and this is being done with such enthusiasm as to threaten the existence of some species. The Oncocyclus Iris are relict populations found in restricted terrains in relatively small communities and attention should be given to their preservation.

It will be noted that 'Capitola' includes two Oncocyclus species, *gatesii* from 'William Mohr' and *iberica* from 'Ib-Mac' (although this latter might prove to be the very similar *elegantissima*). As a result its descendants show more similarity to the Oncocyclus than the descendants of 'William Mohr'. Most of the later Oncobreds show a chromosome count of 47 (or possibly 46). Most modern breeders nowadays cross cultivars together, apart from those breeders who are crossing cultivars with Oncocyclus species. This is still giving splendid results, but more spectacular results may be expected from the attention given of recent years to the breeding of dwarf and the so-called median Iris. These have been crossed with the Tall-Bearded and, as we shall see, have thereby introduced numerous new species and with them numerous fresh genes.

The dwarf bearded Irises of southern Europe have formed the basis of some fascinating speculations by Mitra and Randolph. It is perhaps rather unfair to call them speculations as Mitra's karyotype studies* lend considerable certainty to his conclusions. We start off with two dwarfs with 16 chromosomes, *Iris attica,* which is confined to Greece and *I. pseudopumila* found from Southern Italy to Yugoslavia. At some period when their habitats overlapped these two species must have hybridized

* *Garden Irises,* ed. L. F. Randolph. American Iris Society, 1959.

and the hybrid doubled its chromosomes to produce *Iris pumila* with 32 chromosomes.

At one time it was thought that *I. pumila* was a tetraploid of *I. attica,* but microscopic examination has shown that *I. pumila* contains the chromosomes of both *I. attica* and *I. pseudopumila. I. pumila* has a much more extensive range than either of its parents and is found over much of central Europe. Some strains from Rumania and the Crimea appear to have shed 2 chromosomes and have a count of 30.

There are other dwarf, stemless, Irises in the Balkans and these have 24 chromosomes. They are *II. mellita, reichenbachii, balkana* and *bosniaca.* There is also a 48-chromosome form of *I. balkana,* but this seems to be a tetraploid form of a hybrid between the diploids *balkana* and *mellita,* and should be given a specific name. It will be appreciated that these counts are the same as those of the diploid Tall Bearded and therefore easier to breed from than the 16- and 32-chromosome group.

Finally there is *I. chamaeiris* with a much more extensive range from Spain eastwards and a count of 40. This appears to contain the chromosomes of *I. pallida* and *I. pseudopumila* and seems to have arisen, like *I. pumila,* as a tetraploid hybrid between two species.

Quite distinct from all the preceding, but sharing a dwarf habit with them, is *Iris aphylla* with 48 chromosomes and a range from the Balkans to the Middle East. Morphologically similar but with only 24 chromosomes is *Iris perrieri,* which has only been found in one locality in the Savoy Alps, far from the nearest station for *I. aphylla.*

The earliest dwarf hybrids were interspecific. Almost all the dwarf species will cross, although many of the resultant hybrids are sterile or reluctant to produce viable seed or pollen, although they may do so occasionally. Interspecific hybrids between species with the same chromosome count are highly fertile. A very promising line has been opened by crossing tetraploid Tall Bearded Irises with *I. pumila.* The resultant plants have 40 chromosomes and will breed with the 40-chromosome *Iris chamaeiris.* It has been possible to breed from hybrids with 36 or 44 chromosomes. If *I. chamaeiris* (40) is crossed with *I. pumila* (32), the resultant plants contain 36 chromosomes, while if *I. chamaeiris* (40) is crossed with a tetraploid Tall Bearded (48), the resultant plants have 44 chromosomes. Now if these are crossed with *I. pumila* (32) we get 34 and 36 chromosomes in the progeny. In the next generation of 34 X *pumila,* the odd chromosome drops out and we get a 32-chromosome plant and this can be obtained in an extra generation from the 44-chromosome plant. The resultant plants, although dwarf, show many of the features of the Tall Bearded Irises although pink colour and plicata markings are only transmitted if a white form of *pumila* can be obtained.

An unusual species to be found in some of the dwarf hybrids is *Iris flavissima* the only European Regelia Iris. This has the unusual chromosome count of 22 so that hybrids with *pumila* and *chamaeiris* give 27 and 31 chromosomes respectively. In spite of this, it has been possible to obtain further progeny both by backcrossing to one or other of the parents and also by crossing with Tall Bearded. Why this should take place is rather baffling.

About 1945 American breeders started using *I. pumila* extensively in crosses with Tall Bearded Irises. The pioneer appears to have been Paul Cook who also crossed *pumila* with *flavissima.* His main line of breeding seems to have been crosses between *I. pumila* and the Tall Bearded. The resultant plants have 40 chromosomes and will therefore cross happily with the various forms of *I. chamaeiris* and are also fertile among themselves, although, naturally a certain amount of segregation, but less than one would expect, takes place in the F2 generation.

*Iris aphylla* and *I. balkana* (48) have not hitherto been used with quite so much frequency as *II. pumila* and *chamaeiris.* As so often happens one finds that Sir Michael Foster was a pioneer here, as well. 'Blue Boy' is known to have been an *aphylla* seedling, although the other parent is unknown. There is evidently still plenty to be done in this field.

A very interesting hybrid, raised by G. W. Darby is 'Chancelot' the result of crossing *I. attica* with the pollen of the Tall

Plate III
Bronze-coloured Tree Paeony

Plate IV

*Camellia japonica*

Bearded 'Golden Hind'. *I. attica* has 16 chromosomes and 'Golden Hind' 48, so that the resultant hybrid had 32. However the chromosomes behave rather irregularly. One would expect it to cross happily with *I. pumila* with its 32 chromosomes, but this does not take place, although some seeds were obtained from a *pumila* cross. What occurs is that the 'Golden Hind' chromosomes pair with each other, but the 8 *attica* chromosomes will normally not pair among themselves and will only very occasionally combine with the 'Golden Hind' chromosomes. It will be remembered that *I. pumila* is a tetraploid and so are most of the Tall Bearded Irises so that when these are crossed the *pumila* chromosomes pair with each other and so do the Tall Bearded chromosomes. Hybrids with two sets of chromosomes from each parent (amphidiploids) are generally highly fertile, but the crossing of a tetraploid with a diploid, even though the resultant number may seem promising, is liable to be more or less sterile.

Mention has been made of the Tall Bearded X Oncocyclus crosses and the use of a Regelia in the dwarf breeding. Other Regelias have also been crossed with the Tall Bearded Irises. Of the Regelias, *II. hoogiana* and *stolonifera* are tetraploids with 44 chromosomes, whereas those others that have been counted are diploids with 22 chromosomes. Crosses with the diploid species tend to be highly infertile, although occasionally an unreduced egg cell will give some seeds. On the other hand hybrids with the two tetraploid species are fertile and can be used in second and third generation crosses. The same fertility is reported from crosses with the tetraploid dwarfs *II. pumila, aphylla* and *balkana.* Crosses between the diploid Regelias and Oncocyclus (the so-called Regelia-Cyclus Irises) have long been known and in spite of having 21 chromosomes, they are reasonably fertile and some have been further crossed with Oncocyclus species ('Persian Bronze', 'Persian Lace', etc.) to produce plants with the attraction of the Oncocyclus but the easier cultivation of the Regelias.

A surprising cross is that of various Tall Bearded Iris with *Iris tectorum* of the *Evansia* section of *Iris.* These are the so-called Crested Irises and have a crest where the bearded Irises have their beards. Here again the pioneer work was done by Dykes and Foster. *Iris tectorum* is a native of China, although cultivated for so long in Japan that most works refer to it as a Japanese native. It has a large handsome, rather flat, purple flower. Most of the original Dykes and Foster crosses were lost, but further crossing was done subsequently, although all the plants were more or less sterile. *I. tectorum* has 28 chromosomes. However, recently Mr G. W. Darby obtained through colchicine treatment a tetraploid form of this Iris which should produce amphidiploid hybrids with the Tall Bearded Irises, these should be fertile. The only other hybrid recorded between bearded and beardless Irises is a cross between. *I. pallida* and *I. verna*. *Iris verna* is a dwarf early-flowering North American species that has a series to itself, as it resembles no others. It has 42 chromosomes, so presumably the hybrid has 33 and is highly unfertile. (See also p. 248.)

After the various bearded Irises, it is with the Japanese *Iris kaempferi* that the plant breeder has had the most spectacular success. Although native to various parts of eastern Asia, it is in Japan that the breeding has been done and it is from Japanese plants that all European and North American plants derive. Those who read Japanese Floras and gardening books may be surprised to see that they do not refer to *I. kaempferi* at all, but to *I. ensata*. According to western botanists, *I. ensata* Thun. is a small-flowered species found throughout Asia from the U.S.S.R. to China. It was at one time assumed that it was a native of Japan as well, owing to the fact that Thunberg included it in his *Flora Japonica* (1784), although it appeared there under the name of *I. graminea*, receiving the name of *ensata* in 1786 when it was published in the Proceedings of the Linnaean Society for that year. It is only in comparatively recent times that modern Japanese Floras have been available in translation and it then became apparent that the plant that western botanists refer to as *I. ensata* was not native to Japan at all. Thunberg's description of his *I. ensata* is so sketchy that it might be any Iris. Since he also included the European hybrid

*I. squalens* among the Japanese Flora, one has tended to take his enumeration of species with some reserve.

It would appear to have been Klatt of Hamburg in *Linnaea* in 1867 who first assumed that the plants known as *II. lactea, fragrans, glumacea,* etc., were synonymous with Thunberg's *I. ensata* and he was followed in this by W. R. Dykes in his would-be definitive monograph* on the genus. It seems impossible that *I. kaempferi* could have been derived from this grassy-leaved, small-flowered Iris and no one in the west could understand why the Japanese should claim that *I. kaempferi* was really *I. ensata.* The explanation is quite simple. In the 1920s Dr Bungo Miyazawa visited Uppsala, where Thunberg's herbarium is preserved, and decided that the type specimen of *I. ensata* was the plant known to the west as *I. kaempferi.* His paper on this subject was, presumably, not translated, but it convinced the Japanese botanists. It does not seem to have occurred to any western botanist to inspect Thunberg's specimen until 1967, when at the instigation of Mrs Angela Marchant and the author, Thunberg's specimen was received at Cambridge and inspected by Mrs Marchant and Dr P. F. Yeo. The specimen was well preserved and was undoubtedly the wild form of *I. kaempferi.* The Japanese have proved their point and the correct name for *I. kaempferi* must be allowed to be *I. ensata.* The western *I. ensata* will become, presumably, either *I. lactea* Pallas. or, if this is treated as specifically distinct, *I. biglumig* Vahl. However, for the sake of this discussion it is simplest to continue using the incorrect, but better known name of *I. kaempferi.*

Many writers have stated that *Iris laevigata* has been bred into *I. kaempferi,* but others have thrown doubts on this statement and the latest opinion seems to be that the various forms are all the result of selection and interbreeding. Although they have been cultivated for at least five hundred years the first breeder whose name is known to us is Sho-oh (or Showo) Matsudaira, who flourished around 1840 and introduced two hundred named varieties. The majority of these are known

* *The Genus Iris* (Cambridge, 1913).

as Edo Irises (Edo is the old name for Tokyo). Some of the best of Sho-oh's cultivars were bred in the Higo district and these now form a separate strain. A third strain started at about the same time has been named the Ise strain, from the region in which they were first raised. The Edo strain has no special characteristics, but both the Higo and Ise strain have.

For the Higo breeder the style was the characteristic feature. Flowers with small styles were viewed with disapprobation and the ideal was a broad style, obliquely opened upwards with a large petal-like crest at the top of each style. Varieties with three falls should have these overlapping at the base and slightly concave, while the standards should not be narrow nor too erect. In those varieties where the standards are replaced by three additional falls, the same convex shape was required. In some cultivars the styles become completely petaloid, so that a plant may have nine or more falls, but this was frowned upon by the classical Higo breeders, although modern double Higos have been bred. The original Higos were bred for pot culture and were exhibited indoors. As a result the colours were confined to those regarded as most appropriate for indoor display, white, blue and deep purple; the stems were somewhat weak as the plants were not expected to flower out of doors, so that when they were eventually planted out they proved unsatisfactory. However this disadvantage is said to have been done away with in the modern varieties. The plants were grown indoors in order to display the special nature of the flower which changes in a noticeable manner after it has opened. By the second day it has nearly doubled in size and the flower is supposed to be at its best at midnight on the second day after opening. Modern Higo Irises are occasionally double or treble, which the old Higo never were.

The Ise strain is even more severe and only single flowers are considered tolerable. On the other hand the colours of the Ise strain are very delicate in varying shades of pink and pale blue and other pastel shades. Ise Irises have longer leaves than those of the Higo strain. They are less spectacular than the Higos, but more satisfactory as garden plants. The original varieties had rather thin petals, but the modern Ise strain has strong and weather-resistant petals.

The Edo strain has no special characteristics and includes some old cultivars which differ markedly from the general idea of this Iris. Thus 'Gyoko-Houren' has the falls upright so that the flower looks like a ball, while the twisted narrow upright petals of 'Ryo-No-Tsume' resemble a dragon's claw. 'Ten-Nyo-Kan' has twelve small horizontal falls and looks like *Clematis sieboldii.*

Although *kaempferi* Iris are planted in moist situations in the west, there is apparently no reason for this and the native plants grow on dry hillsides. When they are cultivated in Japan the ground is frequently flooded when they are in flower, in order to give a more decorative effect, but once the flowering is over the land is immediately drained again.

The breeding of both bearded and *kaempferi* Irises takes place with various forms, that are currently regarded as desirable, in mind. One wonders whether many attractive flowers are not distributed because they do not conform with the present rules. It is desirable for a *kaempferi* Iris to look like a clematis, but reflexing standards in a bearded Iris would be thought of as a blemish. It is a feature of all man-made flowers that they have to conform to current trends. If the feathering of tulips had not been virus-caused, the plain coloured tulips would have been lost. As we shall see, when the pansy was first bred, the blotched forms that are the most popular nowadays were completely disregarded and it was only the fact that Continental growers liked them that prevented their disappearance. The prizes of the show bench are valuable as an incentive to flower breeders, but one does sometimes wonder how many desirable plants are lost, because they offend against the rules.

Although all the other sections of Iris have undergone hybridization and selection, the deviation from natural wild plants has not been marked. The most popular are those of the *Xiphium* series (which Dr Rodionenko* would put in a separate genus). These are derived from plants found in North Africa

* In the *Journal* of the British Iris Society, 1962.

and the Iberian peninsula. They are popularly known as Spanish, English and Dutch Irises.

Of these the English Iris, *I. xiphoides*, has the most restricted range in the wild, being confined to the Pyrenees and the mountains of Asturias. With 42 chromosomes, *I. xiphoides* has not been known to cross with any other species, although a cultivated form of *I. tingitana* with 42 chromosomes exists. However, since *I. tingitana* normally has 28 chromosomes, the 42-chromosome form must be a triploid and would be infertile. The name English Iris was given to this Pyrenaean native, owing to the fact that it was an early inhabitant of English gardens: it is said to have been introduced in 1571. In the last century it was given many cultivar names, though these now seem to have been lost. The *Floricultural Cabinet* for 1842 lists no fewer than 76 named cultivars, to which something like another 56 can be added from an earlier number. It is curious that in cultivation all *xiphoides* acquire a mottling of the petals, which suggests a virus, but this has not been proved. In the 1840s there appears to have been a wider range of colours than is available nowadays. One would like to see 'Grand Protector' with a 'rose ground, distinctly margined and spotted with crimson' or 'Flora' with an 'extra carmine flake on a white ground'. Yellow is not a colour that is associated with English Iris, yet both 'Magnet' and 'Mirage' were described as white and yellow, while a number of cultivars including 'Adonis', 'Henriette', 'La Brillante' and 'Sophia superba' were a mixture of blue, white and yellow. I do not know if the number of French cultivar names were the result of snobisme or indicate French raisers. The references to crimson colours suggest the influence of *I. filifolia,* in which pink forms are not uncommon. However, this species was not introduced to British gardens until 1869 so it seems improbable that it could have contributed its crimson colour to these cultivars.

The Spanish Irises are cultivars of *Iris xiphium* ($2n=34$), a somewhat variable species found throughout most of the Iberian peninsula and in North Africa, where may be found also the so-called white-flowered var. *battandieri* which has 36 chromosomes. An early flowering form has been given the varietal epithet *praecox* and a yellow form found mainly in Portugal has sometimes been treated as a separate species, but is now regarded as var. *lusitanica.* These are still generally offered as named cvs. and again there seems to be less variety than in the 1840s. Among plants offered in those days were many with green in the petals: 'Cato' was white and green, 'Christine' was purple, green and yellow, 'Dorothea' was purple and green. Two cultivars, 'Favourite' and 'Hortensia' were brown and yellow. Besides this the colours that are available nowadays were available then, including the bronze. All this must have been due to selection, though presumably the breeding became slightly more controlled as the century progressed.

By comparison the Dutch Iris, which are all hybrids and mainly sterile, are comparatively modern. Most of them have *I. xiphium praecox* as one parent and some are hybrids of this with *filifolia* and *juncea,* while the popular 'Wedgewood' is a hybrid with *I. tingitana.* Foster created hybrids between *juncea* and *filifolia,* a hybrid that might be repeated and further generations bred since as both species have the same chromosome count, (32) the hybrid might be expected to be fertile. *Iris boissieri* a purple *Xiphium* with a rudimentary beard has been crossed with *juncea* and with *tingitana,* but the hybrids are probably sterile and have been taken no further.

A race of Irises that is practically unknown to the British is that known as the Louisiana Irises. These are now thought to be basically hybrids of three species all found in Louisiana, two being marsh dwellers, while the third grows in dry pastures. All three species have been hybridizing in the wild over a long period and, when interest was first aroused, the botanist J. K. Small between 1927-9 described seventy forms to which he gave specific rank. The taxonomic confusion was cleared up by Viosca in the American Iris Society *Bulletin* for April 1935. The three basic species are *II. fulva, brevicaulis* and *giganticaerulea.* To this must be added a colony of very tall reddish Irises found in a cypress swamp near Abbeville, Louisiana which appear to be somewhat distinct from any of these, although most closely resembling *fulva.*

*Iris fulva,* itself, is copper-coloured typically, but the range spreads to yellow and intervening shades. The standards droop slightly giving the flower something of the appearance of a small *kaempferi.* In its natural habitat it is found from New Orleans northward to southern Missouri in wet places, often in depressions that are flooded and often in partly shaded positions. The 'Abbeville' Iris is similar, but larger in all its parts and is subject to deeper shade and more extensive flooding.

*Iris giganticaerulea* (*I. hexagona* var. *giganticaerulea*) is a very robust inhabitant of the coastal marshes between New Orleans and the Texan border. Growing from 3 to 6 feet high, it is typically lavender-blue, but albinos are not uncommon. It thrives under flooding during the winter and spring and will tolerate brackish water that will kill other Iris species.

*Iris brevicaulis* (*foliosa*) is a smaller species than the other two and is a habitant of dry pastures and uplands. As in the case of *I. giganticaerulea* the standards are upright, which distinguishes these two from *I. fulva* quite apart from the colour. The colour of *I. brevicaulis* is blue, but again white forms are not uncommon. *I. brevicaulis* flowers later than the other two species. Both *I. giganticaerulea* and *I. brevicaulis* seem to exist in 42- and 44-chromosome forms, whereas *I. fulva* is only reported with 42. However, not only do they interbreed in the wild, but the hybrids are themselves fertile. The colour range of the natural hybrids ranges from purple, through blue, copper red, pink, coral and flesh to yellow and white and controlled breeding has produced an even wider range. One collected form of *I. giganticaerulea* has been found to be a triploid and breeders are now trying to induce polyploidy.

Although they survive unhappily in Great Britain, the Louisiana Irises need a warmer climate than ours; it is a pity that so attractive a race should be denied our gardens.

The series *Californicae,* the Pacific Coast Irises of North America are far easier to cultivate in this country, but, owing to the fact that they are very difficult to transplant successfully, once the seedling stage has been passed, breeders have not been

19. *Narcissus pseudonarcissus* 'Hispanica', 1810

20. *Narcissus moschatus*, 1810

21. *Narcissus poeticus* 'Ornatus'

22. *Narcissus poeticus* 'Radiiflorus', 1792

23. *Pelargonium scandens*

24. *Pelargonium frutetorum*

25. *Pelargonium inquinans*

26. *Pelargonium zonale*

27. *Fuchsia triphylla*

28. *Fuchsia cordifolia*

29. Seedling Fuchsias

30. *Fuchsia denticulata*

31. Fuchsia 'Dominiana'

32. *Viola altaica*

33. Show Pansies

34. *Viola suavis*

35. *Begonia pearcei*

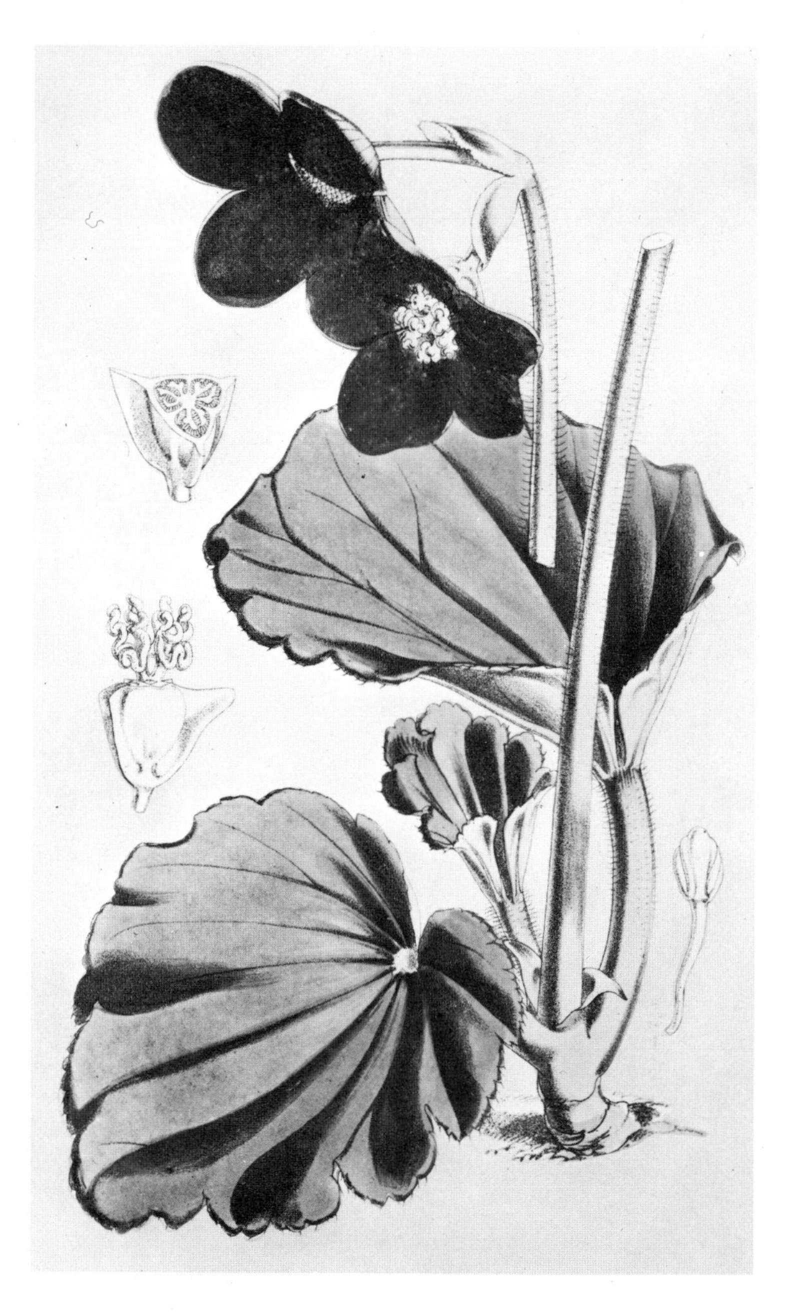

36. *Begonia veitchii*

37. *Bouvardia longiflora*

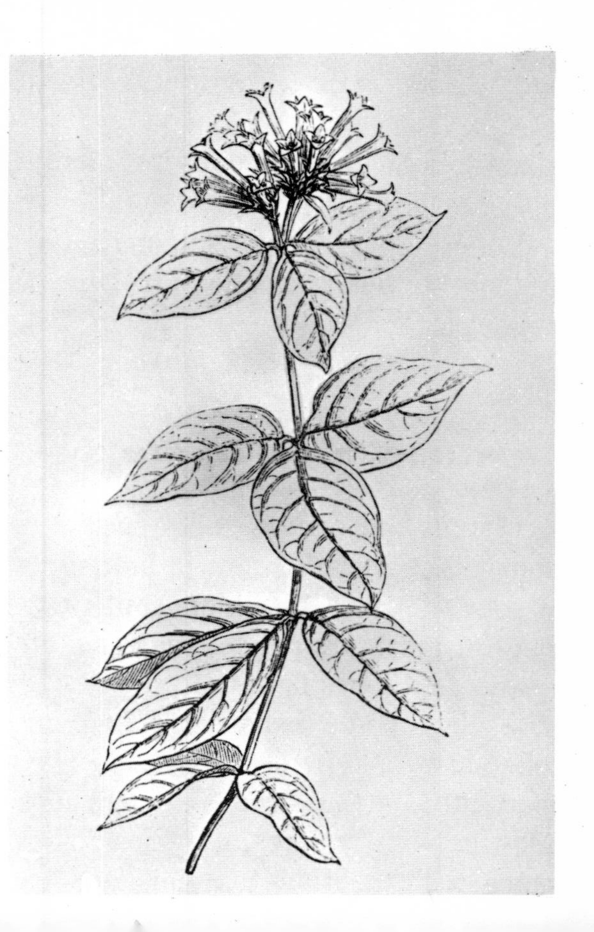

38. *Bouvardia leiantha*

39. *Calceolarias*

40. *Canna glauca*, 1835

encouraged to give cultivar names and it is, in any case, difficult to increase any clone. The *Californicae*, at the latest revision, contains twelve species all of which have 40 chromosomes and all of which appear inter-fertile. Indeed when two or more species occur in the wild in proximity, hybrids are liable to be formed in nature. The principal species that are used in hybridizing are *II. douglasiana, tenax* and *innominata. II. munzii* and *bracteata* are also used, although to a lesser extent. *Iris munzii*, although discovered comparatively recently, is a very desirable species with large flowers of a very fine blue, and it transmits its large flowers and fine colour to its progeny. Unfortunately it is rather frost-tender and its use is thus restricted to mild areas. *Iris bracteata* is rare in the wild, being confined to a single area in Oregon, where it grows in dry parts of forests. It is even more resentful of root disturbance than the other *Californicae* and this has made its use comparatively restricted. This seems a pity as it has very lovely flowers which are golden-yellow, veined with dark brown or dark purple lines.

*Iris innominata* is a smaller plant, only discovered in 1928, with grassy leaves and comparatively large flowers. These are usually some form of yellow, but lavender and purple forms have been found, although at one time these were given the name of *I. thompsonii* (this name now covers natural hybrids between *innominata and douglasiana*). *I. innominata* has been used a lot in controlled hybridizing, principally with the very variable *I. douglasiana.*

*Iris tenax* was the first of this series to be introduced into cultivation. It was collected by David Douglas in 1825. It is larger than *I. innominata,* but the flowers cover the same colour range. At one time the yellow form was called *I. gormanii.* It has the most northerly range of the series and is one of the easiest to grow.

*I. douglasiana* is the most well-known and the most variable of this series. Its range covers a length of seven hundred miles, but it is always found near the coast. In height it varies from 12 to 36 inches and in colour from pale cream to lavender and deep purple. Sky-blue and white forms have occasionally been found. This is the species most frequently used for hybridizing and with its inherent variability is obviously a most valuable parent. The Californian hybrids are liable to remain in a state of flux, owing to the difficulty of propagating clones vegetatively. The *Californicae* show affinities with the East Asian *Sibiricae* and successful crosses have been made not only with the 40-chromosome species, but also with the two 28-chromosome species *sibirica* and *sanguinea* (*orientalis*). These hybrids are very attractive, but completely sterile. Evidently the chromosomes are too distinct to pair, even when the diploid count is identical, so there seems no chance of advanced generation hybrids.

The *Sibiricae* themselves have been inter-hybridized, but most of the so-called Siberian Irises are crosses between *I. sibirica* and *I. sanguinea* and are not so very different from the species. However there are white varieties and at least one that is magenta. Recently breeders have been crossing the 28-chromosome group with the two yellow species, *wilsonii* and *forrestii* in the 40-chromosome group. This has been effected successfully but the resultant hybrids, though very attractive, are, not unexpectedly, sterile. More spectacular results may be expected from crossing the two yellow species with the tall dark blue *I. delavayi,* as this belongs to the 40-chromosome group and the offspring should be interfertile. *I. sibirica* has been crossed with Irises from other series such as *II. pseudacorus, laevigata, setosa* and *versicolor*, but such crosses have all proved sterile and do not seem to have entered into commerce. There appears to be no record of polyploidy in these Irises, so that the chance of producing amphidiploids is remote.

In Japan selected forms of the water-loving *Iris laevigata* have been cultivated. At one time nearly a hundred cultivars were available, but now there are only about ten. These are coloured white, blue, purple and violet. Some double forms are known and the cv. 'Mai-Kujaku' has eight petals. These are not known in the west, but it is possible that they may have occasionally been included in shipments of *I. kaempferi* and so given rise to the belief that this was a factor in the development of the *kaempferi* cvs. There is no reason why they should not be grown over here as the species has long been a garden favourite.

Certain series still await the attention of the hybridizer, notably the confusing *Spuria* group, which contains some very fine garden plants. It would seem improbable that these would give such spectacular results as have been achieved in the bearded Irises and in the *kaempferi*, but when one considers that this latter has been developed without any admixture of any other species, so far as is known, it is hard not to be optimistic.

CHAPTER VIII

# Narcissus

Although Narcissus have been favourite garden plants for well over three hundred years, it is only comparatively recently that they have been the subject of deliberate hybridization. They have, however, for long hybridized in the wild in the lands bordering on the Mediterranean where they are native.

In 1629 Parkinson described ninety different sorts of Narcissus as being cultivated in gardens but interest in the flower seems to have waned subsequently and when Miller came to write his *Gardener's Dictionary* in the middle of the eighteenth century he listed only nine species, two of which, *NN biflorus* and *incomparabilis*, we now know are natural hybrids. Among these Miller mentions five varieties of the trumpet daffodil and also the double white pheasant's eye, but the species which seems to have been in most demand was *N. tazetta.* He lists ten named cultivars of the form with a white perianth and orange cup and twelve with yellow perianths and orange cups. It is interesting to see among these, names that are still familiar today. 'Grand Monarque' is found in the first section and 'Soleil d'Or', englished as 'The Golden Sun' among the second.

By 1839, in his *Hortus Britannicus* Loudon was listing seventy-six species, some of which are now unidentifiable.

The first deliberate hybridizations were undertaken by Dean Herbert in the 1820s to prove his thesis that certain wild plants were not species but were natural hybrids. The most important of these, which has yet to be mentioned is *N.* X *incomparabilis* (*N. pseudonarcissus hispanicus* X *N. poeticus*). By 1843 Herbert named some of his hybrids. 'Diomedes' was *asturiensis* (*minimus*) X *tazetta*; 'Pallidus' was *pseudonarcissus minor* X *moschatus*; 'Spofforthiae' was *incomparabilis* X *poeticus* (that is a second injection of poeticus blood into a poeticus hybrid, but Herbert used this time the form of *poeticus* known as *stellaris*); and 'Sub-concolor' which was *asturiensis X poeticus*.*

Herbert's experiments encouraged others to follow his example, the most noteworthy being Edward Leeds (1802-77) and William Backhouse (1807-89). Their collections were bought, on their deaths, by the nurseryman Peter Barr, who developed their work further. He was joined in this by Mr and Mrs R. O. Backhouse, who wished to develop a daffodil with a pink trumpet and also by the Reverend George Engleheart (1851-1936), possibly the greatest of all hybridists in this field. He worked especially with the poeticus group.

Edward Leeds was interested in raising white or very pale flowers. 'Leedsii' is *N. montanus* X *N. poeticus*. It is not quite clear if by *N. montanus*, Leeds was referring to *N. moschatus* of which *montanus* had been a synonym, or whether he was referring to *moschatus* X *poeticus*, which is now known as *N.* X *boutignyanus*, but which has also been termed *montanus*. At one time there was a race of daffodils known as the Leedsii race and these were all crosses of *poeticus* with trumpet daffodils. One cross of Leeds's, which was somewhat of a novelty was *N. hispanicus* X *N. calathinus*. This last name is a now superseded one for the giant form of *N. triandrus* found on the Ile de Glenans off South West Brittany, far away from all other forms of this plant, which is confined mainly to North-West Spain. It was formerly known as var. *loiseleurii*, but Dr Fernandes now considers it as the same as the Spanish var. *cernuus*. It is this giant French form which is employed in most of the *triandrus* hybrids.

Before going further into the hybridization of the genus, it might perhaps be worth while to consider the various sections into which Dr A. Fernandes has divided the species.† Of his sixteen sections, seven are devoted to natural hybrids between the various sections, so we can conveniently regard Narcissus

* *Botanical Register*, No. 35 (London, 1843).

† In his revision of the genus in *The Daffodil and Tulip Year Book* (London, 1968).

proper as divisible into nine sections. Among these are some plants that are very rare in cultivation and have not been used for hybridizing. The section Aurelia consists of a single species *N. broussonetii*, which is obviously related to the Tazetta group (Section Hermione) but which has practically no corona. The plant is native to Morocco and requires greenhouse treatment in this country. It has a chromosome count of $2n=22$, which is paralleled among most of the yellow forms of *N. tazetta* (*NN bertolonii* and *aureus*) and also many others. *N. broussonetii* could presumably hybridize with some of the yellow tazettas, which might give us a rather tender yellow, autumn-flowering plant, with a head of some 12 or more flowers; a plant that could well be useful.

For most of the species the haploid count is either 7 or 10. An exception is *N. serotinus* (Serotini section) which has been found with 10 or 30 chromosomes, suggesting that there must have been an ancestral species with a haploid count of 5 and that the plants in the section Hermione with the count $2n=20$ must be regarded as tetraploids, not diploids. *N. serotinus* is believed to have hybridized in the wild with a Hermione, *N. bertolonii*; the resultant *N.* X *chevassutii* has been found in Algeria. No similar work has been done in cultivation, probably because *N. serotinus* is by no means easy to cultivate.

The section Hermione contains all the tazetta group. Some of these have $2n=20$ and others $2n=22$. They all have a number of flowers to the scape and these may be all white, yellow with an orange cup or white with an orange cup. The plants have a circum-Mediterranean range and have hybridized in the wild with species from other sections. Such hybrids are usually sterile, but others, in spite of their odd chromosome count, $2n=17$, often appear to be fertile. *N.* X *biflorus* (*medio-luteus*) a hybrid between *poeticus* and *tazetta* has been introduced and naturalized into at least seven countries.

The section Narcissus contains all the species that were formerly regarded as subspecies of *N. poeticus*, but which are now divided into *N. poeticus* and *N. radiiflorus*. They are found on the mountains of Europe from Spain to Greece and all have one-flowered scapes with white perianths and a small cup, which contains varying amounts of a red colour. The haploid number is 7 but some species are triploid. This section has been much used in hybridization, particularly *N. radiiflorus* var. *poetarum*, which has a wholly red cup.

The Section Jonquilla with $2n=14$ contains a number of rather dwarf plants with 2 to 5 yellow flowers on the scape and usually a pleasant fragrance. The section has been used in hybrids, but these usually appear to be sterile and so have not proved susceptible to further crossing. The species are based on the Iberian peninsula, but just get into southern France and the somewhat aberrant, autumn-flowering *N. viridiflorus* ($2n=28$) is confined to Gibraltar and Morocco.

The section Apodanthae consists of dwarf plants, generally single-flowered with a round perianth and a cup-like corona. They all appear to be $2n=14$. They are found in the Iberian peninsula and Morocco and are all various shades of yellow, with the exception of the white *N. watieri* from the Atlas. A natural hybrid between *N. scaberulus* and *N. triandrus* has been recorded in Portugal. The section has not been much used in commercial hybridization, but has been used for the raising of miniatures.

The section Ganymedes contains two species, the yellow *N. concolor* and the creamy-white *N. triandrus*. The scape bears from one to five flowers with a cup-shaped corona, which is quite deep and slightly reflexed perianth segments. It has been used quite extensively in hybridization. $2n=14$.

The section Bulbocodium (Corbularia in earlier authors) has a great variation in its chromosome count: $2n=14$, 21, 28, 35, 42, 49, 56 and it is probably this that accounts for the fact that it has been little used in commercial breeding, although a number of wild hybrids have been noted. The plants are generally rather dwarf with a single-flowered scape. The flower has a very large spreading trumpet and very narrow perianth segments. The range is from South-West France to Morocco and the flowers are yellow except for two whitish species *N. romieuxii* from Morocco and *N. cantabricus* from Morocco and southern Spain.

The section Pseudonarcissus (Ajax in older authors) contains all the trumpet daffodils. These usually bear only a single flower to the scape (*N. longispathus* is said to bear two occasionally) and are either entirely yellow, bicoloured with a whitish perianth and a yellow trumpet, or cream-coloured (*NN. moschatus* and *alpestris*). Dr Fernandes includes in this section the dwarf, yellow, *N. cyclamineus* with very characteristic reflexed perianth segments and a habit of very early flowering. The majority of this section is $2n = 14$. The bulk of the species are native to the Iberian peninsula, but *N. pseudonarcissus* is found in most European countries and it is hard to believe that all occurrences are due to naturalization. This section is the one that has been used most extensively and there are few large-flowered hybrids which do not contain some Pseudonarcissus. The only exception is to be found in the Poeticus hybrids. The first of these, Engleheart's 'Horace' was a cross between *N. radiiflorus* var. *ornatus* and *N. radiiflorus* var. *poetarum* (it has been supplanted by 'Actaea', which must have similar parentage).

The trumpet daffodils are all crosses of the various species. The yellow daffodils are usually based on *N. hispanicus*, the bicolors on *N. bicolor* and the white trumpets on either *N. moschatus* or *N. alpestris*, though two other whitish species *NN. tortuosus* and *albescens* have also been used. Bicolors can be formed by crossing the white species with the yellow ones.

The large-cupped hybrids are crosses between trumpet daffodils and the section Narcissus. The red-cupped forms had *N. radiiflorus* var. *poetarum* as one parent. One of the first of these, which was much used in subsequent breeding was Engleheart's 'Will Scarlet' *N. abscissus* (a rather short-trumpeted Pseudonarcissus) X *N. radiiflorus* var. *poetarum*. This has a good red cup, but the perianth segments are starry and rather twisted and this is looked on with disfavour by breeders. Unfortunately this habit tends to be transmitted to its offspring. A better ancestor for red cups and trumpets has proved to be Leeds's 'Beacon'. The parentage of this is thought to be

'Princess Mary' X *radiiflorus poetarum*. 'Princess Mary' was probably *poeticus* var. *hellenicus* crossed with an unknown trumpet daffodil. The red and pink trumpeted daffodils are the result of breeding a Pseudonarcissus with *radiiflorus poetarum* and using the pollen of this cross on the Pseudonarcissus. Small-cupped hybrids are generally the result of crossing a Pseudo-narcissus with a Narcissus and then pollinating this cross with a Narcissus, but they can also be formed with a trumpet daffodil-poeticus cross, using the poeticus as a pollen parent. *N. cyclamineus* has also been much used, particularly with trumpet daffodils, to give such favourite cultivars as 'Peeping Tom' and 'February Gold'. These have the trumpets of the large cvs. and the typically reflexed perianths of *N. cyclamineus* as well as this species' early flowering habit.

The race known as Poetaz is, as might be inferred, the result of hybridizing forms of *N. poeticus* with the pollen of various Hermione species. Many of these latter are somewhat tender in northern climes and the effect of the hybridization is to give multi-flowered scapes, in which the individual flowers are somewhat larger than those in the Hermione section, but with the increased hardiness and later flowering of the Narcissus section. Both *N. radiiflorus* var. *exertus* and var. *poetarum* have been used as seed parents and white and yellow Hermiones as pollen parents. The resultant plants are attractive, but as they are sterile it has not been possible to pursue these crosses for more than one generation.

It would seem, therefore, that Narcissus breeding has got into rather a rut. Apart from the dwarf Apodanthae, there are not many species remaining to be used and breeders tend to be confined to crossing the existing cultivars in the hopes of finding larger and better exemplars. One result of this crossing has been the emergence of a recessive gene which splits the trumpet into petal-like segments, but the plants are curious rather than attractive. The trumpet segments reflex, so that they are nearly parallel with the perianth segments. The race is known at the moment as Split trumpet daffodils.

One possible source of further breeding has been found as

a result of modern cytology. For a long time a natural hybrid between *N. requienii* (*juncifolius*) of the section Jonquilla and the pure white Hermione *N. papyraceus*, has been known and given the name *N.* X *dubius*. Investigation has proved that this has the unusually high chromosome count of $2n=50$. Of these 50 chromosomes, 28 come from *N. requienii* and 22 from *N. papyraceus*. The normal counts of these two species is 14 and 22. What would appear to have happened is that originally a tetraploid form of *N. requienii* ($2n=28$) must have crossed with *N. papyraceus* ($2n=22$) to give a hybrid with $2n=25$. Then a tetraploid form of this must have arisen to give the count of $2n=50$ and, incidentally, to provide a fertile plant. A further hybrid of this with *N. requienii* has been found wild in Catalonia and given the name *N.* X *pujolii*. *N.* X *dubius* gives a pointer as to how further breeding might be carried out among sections which do not show much affinity at the moment. If amphidiploids can be created, it should be possible to employ the section Jonquilla for second generation breeding. Cytology will also help in the use of the section Bulbocodium, although it is not very clear what advantage this will bring. The objection to this course of action is the enormous time it takes to raise Narcissus seedlings. Anything from five to ten years must be reckoned between sowing and flowering. It should be possible to know after a year if tetraploids have been raised by artificial means such as colchicine, but it might be fourteen or more years before the first cross flowered and few breeders have that sort of patience. On the other hand it might well appeal to amateurs and there seems little chance of new breaks without such a method of breeding.

There might also be advantages in reproducing some of the earlier crosses. In an article in the *Gardeners' Chronicle* of June 1865, William Backhouse recorded that among his crosses of *N.* X *incomparabilis* with *N.* X *poeticus* were some plants with greenish flowers. No one wanted green flowers in 1865, but nowadays they might be welcomed and this would seem to be a line worth investigating.

The three autumn-flowering species are rare in cultivation

and not very easy to flower, but there seems no reason why a race of autumn-flowering Narcissi should not be raised. *N. serotinus,* the most widespread is perhaps the least useful as it has so odd a chromosome count, although the natural hybrid *N. bertolonii* X *N. serotinus* sounds an attractive November-flowering plant. On the other hand *N. elegans,* now placed among the Hermiones with $2n=20$ should hybridize successfully with other Hermiones with the same count (*NN. tazetta, patulus* and one form of *bertolonii*) and the resultant hybrids should, with luck, be fertile. *N. viridiflorus* is now placed among the Jonquils. It is a tetraploid with $2n=28$ and in order to get beyond the first generation it would be necessary to have other tetraploid Narcissi to use as other parents. There exist certain members of the Bulbocodium section with this count already and crosses between *N. viridiflorus* and *N. romieuxii* or *N. cantabricus foliosus* are theoretically quite possible and should be fertile. Crosses with tetraploid Hermiones, should such be found should prove very valuable. The unusual green colour of *N. viridiflorus* would obviously be very interesting to introduce. There exists already a tetraploid Jonquilla, *N. fernandesii* var. *major* and this would seem a suitable species with which to start.

A number of dwarf hybrids have been raised such as *N. asturiensis* X *cylamineus* ('Minicycla') and *N. aureus* X *cyclamineus* ('Cyclataz') and a number of *triandrus* hybrids. These are all quite agreeable plants but do not appear to have many future potentialities.

## CHAPTER IX

# Pelargonium

There are three distinct hybrid races of Pelargonium: the so-called Regals, the Zonals and the Ivy-leaved. We know a certain amount about the development of the last two sections, but the Regals appear to be excessively complicated.

*Pelargonium* is a large and complicated genus, which Sweet, one of the first monographers of the *Geraniaceae,* wished to subdivide into eight genera. De Candolle ignored the proposed generic separation, but used many of Sweet's proposed names as subgenera and this classification is maintained, more or less unaltered, to this day. The majority of the probable parents of the Regals (*P.* X *domesticum*) come from the sub-genus *Pelargium* and include *P. cucullatum, angulosum, capitatum, cordifolium* and *betulinum*; possibly *P. scabrum* was also used. The other species are *P. grandiflorum* from the subgenus *Eumorpha* and *P. fulgidum* from *Cortusina.* Of these species *P. grandiflorum* would contribute a white colour and *P. fulgidum* a brilliant scarlet. The others are various shades of purple. We should perhaps consider them in more detail.

*P. angulosum* is a shrubby species, growing to about 3 feet high in the wild, with sharply angled leaves of varying shapes. One form, often regarded as a separate species, is var. *acerifolium.* The flowers are in panicles of 4 to 6 flowers and are carmine-purple.

*P. cucullatum* is similar, but more vigorous and with rounded leaves. The flowers are crimson with marked purple veins.

*P. capitatum* is a smaller plant with dense heads of small rosy-purple flowers. The cordate leaves are perfumed.

*P. cordifolium* is a small, but vigorous plant with a long

flowering season. The upper petals, which are rosy-mauve with dark red veins are much larger than the self-coloured lower petals.

*P. betulinum* is a weak short plant with rarely more than 2 or 3 flowers in the peduncle. The flowers are a pink shade of mauve with heavy crimson veining in the two upper petals.

*P. scabrum* is about 2 feet high with deeply-lobed leaves and purple flowers. One of the earliest hybrids was *scabrum* X *glaucum*. The latter is a member of the subgenus *Campylia* and appears to have been much used in early hybridization, although its attractions are far from obvious.

*P. grandiflorum* is a shrubby plant about 2 feet high, with smooth glaucous leaves and large flowers, which are white with red veins on the upper two petals. The two upper petals are large and wide, while the three lower ones are narrow and widely separated from each other. Like the next species, it would seem that its main contribution to the strain was colour.

All the previous species have had open flowers, but those of *P. fulgidum* have a long tube. The plant is covered with silvery hairs. The flowers are a brilliant scarlet with vermilion veining.

Of all these species, the most significant are *PP. angulosum, cucullatum, grandiflorum* and *fulgidum.* The first two species seem to give the plant its habit, the latter two only appear to have contributed their colours. The various species were extensively hybridized in the early nineteenth century and the hybrids were then crossed and back-crossed, so that the strain is now very involved. The principal hybridizers seem to have been Sir Robert Hoare and the nurseryman Colvill. Crosses between *P. fulgidum* and various scented leaved species, probably principally *P. capitatum,* are sometimes known as the 'Unique' strain. Sweet's *P.* 'ignescens' is probably one of these.

The early hybridists either took no records of their crosses or failed to communicate them. Even Sweet has usually to hazard a probable parentage. In the long run it probably matters little, although it would be nice to have more exact pedigrees. However by now the original species have been so hybridized, that any putative parentage of modern cultivars is out of the question.

By comparison, the Zonals appear more straightforward. These are all hybrids from the subgenus *Ciconium,* which appear incompatible with species from *Pelargium,* but will hybridize amongst themselves and also with the Ivy-leafed *P. peltatum.* The Zonal Pelargonium is usually stated to be *P. inquinans* X *P. zonale,* but, as Derek Clifford points out in his excellent *Pelargoniums* (London, 1958) to which this chapter is heavily indebted, it is probable that the early hybridists had not the true *P. zonale* at all, but either *P. scandens* or *P. frutetorum.* In spite of its name *P. zonale* has not got a heavily marked leaf, while the other two species have very marked leaves. The mark is thinner in *P. scandens,* which is probably one of the ancestors of the so-called Tricolors, as its leaves show the 'butterfly mark' at the base. Other species that may have entered into the strain are *P. salmoneum* and the mysterious *P. monstrum.* In modern times the strap-petalled *P. acetosum* has been used.

Of all these species *P. inquinans* is the most striking with its broad-petalled scarlet flowers. By selection it was possible to breed a flower nearly completely circular, which always appears to have been the nineteenth century ideal shape for a flower. All Pelargoniums have asymmetric flowers and the upper petals remain very slightly larger even in those cultivars that appear to be completely circular and symmetrical.

The species that would seem to cross most happily with *P. inquinans* to make a full-petalled flower with a markedly zoned leaf is *P. frutetorum,* which has a solid zone in the leaf and quite wide-petalled salmon-pink flowers. It is not too certain that it has had any influence on the original Zonals. It was not separated as a species until 1932 and it is not clear as to whether it had been in cultivation before, without its specific identity being realized. At a time when *P. zonale* seems to have covered two or three species it is not impossible that plants may have been grown under this blanket name. There is no doubt that *P. scandens* was in cultivation after 1790 and this has almost certainly played a part in the Zonal hybrids (*P.* X *hortorum*). It makes a straggling, long-jointed shrub with a very distinct zone on the leaf and many-flowered umbels of pale pink flowers, and I would say its influence can certainly be observed in such cvs. as 'Maxime Kovalski'.

*P. zonale* itself has only a very faint zone on the leaf and markedly asymmetric purplish flowers. The two upper petals are close together, while the three lower are markedly separate. The umbel carries many flowers and there is no doubt that it has contributed to the strain, although probably not so much as has been reported. It would be gratifying if someone would repeat a series of crosses between *P. inquinans* and other members of the *Ciconium* group, so that we could see the results.

*P. salmoneum* is a mysterious species that has been in cultivation since 1732, but has not been found in the wild. Linnaeus thought it was a hybrid between *zonale* and *inquinans* and named it *P. hybridum*. It breeds true from seed, so Linnaeus was wrong in his conjecture. *P. hybridum* is the correct botanical name, but would be so confusing in this context that I am using Dyer's synonym. It does look as though *P. zonale* might have some affinity to it, but the leaves are different in shape and unzoned. The umbel is less heavily flowered than that of *P. zonale* and the flowers are a good shade of salmon pink. At some time the plant was lost to cultivation and was not rediscovered until 1929, when it was found in a public garden in Port Elizabeth. It may well have contributed the salmon-pink colour to the Zonal hybrids.

Even more obscure is the dwarf, gnarled *P. monstrum*. This has been known since 1784 and is still in cultivation, but, again, has never been found in the wild. It also breeds true from seed. It has zoned orbicular leaves and rather small pale pink flowers and looks as though it has contributed its dwarf habit to the short ornamental-leaved cultivars.

Although it is impossible to determine how many species are involved in any one cultivar, it is safe to say that the rounded flowers show the influence of *P. inquinans*, while the numerously flowered umbels of the so-called Nosegay types, show the predominance of other *Ciconiums*. The brilliant scarlet colour is only found in *P. inquinans*. There are said to be white and red forms of *P. zonale*, but this species seems insufficiently studied and practically any statement about it should be treated with reserve. It is safe to say that the modern hybrids contain more than the two species to which the strain is attributed.

It is not clear when the first white-flowered cultivar appeared. Peter Grieve in his *A History of Variegated Zonal Pelargoniums* (London, 1868) says that three were known to him as early as 1859, but the first satisfactory plant seems to have been 'Madame Vaucher' raised in France in 1860. Double-flowered cultivars were raised from 1864 onwards by Victor Lemoine and about 1872 the tetraploid strain raised by Messrs Bruant in Poitiers made its first appearance and has influenced many of the larger modern cultivars. The Nosegay strain appears to have dropped from popular favour, although the so-called Cactus-flowered cvs. may be derived from them.

From the mid-1850s the Zonals were given a new lease of life by the breeding of ornamental-leaved cultivars which were known as Tricolors. These were almost entirely the result of the work of Peter Grieve, who was gardener at Culford Hall, Bury St Edmunds. Some pioneer work on raising white-margined Zonals had been undertaken by a certain Kinghorn, whose 'Attraction' is the first recorded silver Tricolor. Grieve started to work with these and produced several improvements, culminating in 'Italia unita'. In 1855 he started to work with the only gold-margined plant known, 'Golden Chain', a plant of whose origin we know nothing. This was crossed with various cultivars of which we have no knowledge nowadays and Kinghorn's 'Attraction' was also used. The pedigree of the famous 'Mrs Pollock' was 'Golden Pheasant' X 'Emperor of the French'. 'Golden Pheasant' was 'Emperor of the French' X 'Golden Tom Thumb'. This latter plant was 'Cottage Maid' X 'Golden Chain', while 'Emperor of the French' was 'Cottage Maid' X 'Attraction', so the end result was a second generation cross from only three cultivars. The breeding of these was taken up with great enthusiasm for some time and in 1868 in the *Floral World* Shirley Hibberd gave the following instructions for raising new cvs.: 'For the production of seeds it is necessary to have a collection comprising a few of the very best varieties . . . , and also a certain number of common zonate kinds, the leaves of which are round, stout, flat or slightly convex and marked with very dark zones. "Madame Vaucher" and "Monsieur Barre" are two of the very best dark zonals for this purpose as their leaves are finely formed and deeply coloured and the plants have vigorous constitutions. The breeder should cross his plants in every way he can imagine, and he will obtain a variety of results, a large majority of the seedling plants proving, as a matter of course, worthless. But in breeding for leaves it will be found that when a dark zonal such as "Madame Vaucher" is made the seed parent, and a golden zonal, such as "Sophia Cusack" is made the pollen parent, a majority of the seedling plants will in due time put forth shoots with golden zonal leaves. In like manner if pollen is taken from a silver zonal to fertilize the flowers of a dark zonal, a majority of the seedlings will be silver zonals. To breed bronze zonals is the most easy task of all, as their own seeds, without artificial fertilization, will produce bronze zonal plants from the first; but another mode of obtaining them is to breed the golden zonals together – as, for example, to apply the pollen of "Sunset" to the stigma of "Yellow Belt". When we say the cultivator should cross every way, it must be understood that he should do so with judgement, and keep in mind always a rule which recent carefully-conducted experiments have confirmed as applicable to the case – that the seed parent usually has most influence on the *form* of the flowers and leaves; and the pollen parent usually has the most influence on the *colour* of the flowers and the leaves. To a certain extent, however, the characters of both parents are usually blended in the offspring; this, also, has been amply proved by long-continued and carefully-contrived experiments.'

I have given this rather long quotation for its interest, not only in the raising of tricolor geraniums, but also for its insight on the theories employed by plant breeders in the mid-nineteenth century. One would like a few details on the 'long-continued and carefully-contrived experiments', but it is probably safe to assume that the remarks on the relative effect of pollen and seed parents was based on observation. It does suggest that cytoplasm may play a larger part in hybridization than some modern writers are prepared to concede and that some characters may be sex-linked.

The Ivy-leaved Pelargoniums are derived principally from *P. peltatum,* which supplies the only species in the subgenus *Dibrachya* (*P. lateripes* was at one time thought to be another species but is now regarded as a hybrid with a Zonal). Members of the subgenus *Ciconium* will hybridize with *P. peltatum* and the Ivy-leaved plants are probably hybrids between these two subgenera, although the influence of the Zonals is only to be found in the varying colours. There have been some reciprocal crosses, such as 'Milky Way' and 'Madame Charmet' where the habit of the plant resembles that of the Zonals, while the flowers show the *peltatum* influence. *P. peltatum* itself closely resembles a rather small-flowered modern Ivy-leaf with mauve or white flowers, the upper petals veined with crimson. The cultivar 'L'Elegante' is, apart from its variegated leaves, very close to the original species. The plant is said to be variable in the wild and many people hold that the older Ivy-leaved are simply selected forms, that have been developed in cultivation. This is considered by Mr Clifford to be unlikely, and I agree with him. We know that an Ivy-leaved X Zonal cross was made as long ago as 1862 and there were probably earlier ones that were unrecorded.

Scented-leaved Pelargoniums are derived principally from *P. graveolens* and *P. crispum. P. graveolens* lends its much dissected leaves to the hybrids, while *P. crispum* and its offspring show an entire, rounded leaf. *P. graveolens* and its related species have been crossed with Regals to give scented-leaved plants with brighter, more showy flowers. As the species are variable and natural hybrids are known, it is by no means easy to identify a number of cultivars and it is probably simpler to lump them together as Scented-leaved Pelargoniums and leave the matter there.

Derek Clifford in his comprehensive volume notes three hybrid strains that have disappeared from cultivation. In Sweet's day there were many hybrids among the subgenus *Hoarea,* which is characterized by tuberous roots and large heads of small, but attractively coloured flowers. In 1895, among his Regals Lemoine listed 44 hybrids with *P. glaucum,* a small-flowered species with long slender glaucous leaves, that might have lent an attractive foliage to the Regals; and in 1880 Cannell listed 8 hybrids of the attractive white-flowered *P. echinatum.* Neither of these strains was persisted with. It may be that although a primary cross was made, the resultant plants were sterile so that no further development was possible, although one cannot imagine Lemoine raising as many as 44 hybrids without finding this out.

Most of the Pelargoniums in cultivation have a basic haploid count of 9 chromosomes. The counts published in Darlington and Ammal's *Chromosome Atlas of Cultivated Plants* are rather unconvincing. Apparently *P. tomentosum* is a pentaploid, while the various Regals counted seem to have either dropped or acquired extra chomosomes on some occasions. Most of the *Ciconiums* are diploid with 18 chromosomes, but there is said to be a tetraploid form of *P. zonale.* However, goodness knows what this represents as it was taken from a cultivated plant, and may well be a Zonal hybrid. *P. peltatum* is listed as a tetraploid. Cytologists are not always botanists and have to depend on other people's identification of species. Was this a wild *P. peltatum* one wonders? As we have said, it is not possible to cross plants from *Ciconium* with plants from *Pelargium. Ciconium* will cross with *P. peltatum* of the subgenus *Dibrachya* and, although the only record is somewhat doubtful, *P. peltatum* might cross with *Pelargium.* If it does, it may eventually be feasible to combine the three strains. Here, at any rate, is a possible line for breeders to follow.

CHAPTER X

# The Fuchsia

*Fuchsia* is a genus of about a hundred species, most of which come from Central and South America, but there is also a strange collection of species from New Zealand. They have a basic chromosome count of $2n=22$ and are mainly diploids, with a few tetraploids known. *Fuchsia triphylla* seems to have a triploid form in cultivation, but this is evidently not invariable, otherwise the plant would be extinct. It would seem, however, that the New Zealand species are not compatible with the American ones, so that hybridization between them has not, so far, proved possible.

It is interesting that the first species to be described was *F. triphylla.* Père Plumier figured it in 1703, but the plant does not seem to have been received into cultivation until around 1872. It was one of the last to be introduced. The first species to be cultivated in Europe were *FF. magellanica* and *coccinea,* which were apparently received at Kew about 1788. *F. coccinea* was distributed in 1793 by James Lee, the famous owner of the Vineyard Nursery, Hammersmith, and was accompanied by a romantic story of how Lee saw the plant growing in the window of a cottage at Wapping. It was said to have been brought back from South America by a sailor and given to his mother. She was unwilling to part with it, but eventually sold it to Lee for eighty guineas, with the proviso that she was to get in addition two rooted cuttings. Lee is reported to have raised three hundred cuttings in one year from this plant and to have sold them at prices ranging from one to twenty guineas, according to which account you read. Unfortunately doubt has been thrown on this picturesque story. In the *Floricultural Cabinet* for 1855, it is suggested that in 1788, a Captain Firth supplied plants of both *F. magellanica* and *F. coccinea* to Kew Gardens and that it was thence that Lee procured his plant, using the story of the sailor's mother to conceal the actual provenance and, also, of course, as a good sales gimmick. The matter is further confused by the suggestion that the plant in question was not *F. coccinea* at all, but the not dissimilar *F. macrostemma,* now regarded as a subspecies of *magellanica.*

It seems rather unlikely that an able seaman would have been able to bring back a plant alive from South America in the conditions under which they worked in 1788, but perfectly possible for a captain to do so, which suggests that the less romantic story is probably correct, alas.

For some time these two species were the only ones in cultivation. A small-flowered plant, *F. lycioides,* was introduced in 1796, but has played no part in developing the genus. The large *F. arborescens* arrived in 1823, a plant known either as *microphylla* or *parviflora* in 1824 and *F. fulgens* in 1830. *F. apetala* is said by Loudon to have been received in 1825, but the usual date of introduction is about 1843, when W. Lobb sent four species *FF. apetala, decussata, dependens* and *serratifolia* to Veitch of Exeter. The showy, but tender, *F. corymbifolia* had come by 1840. The early history of hybridization is somewhat vague. In 1825 the pollen of *arborescens* was used on *F. macrostemma,* but there is no record of the cross being successful. There is also a report of a Fuchsia called 'globosa' (which is a name that crops up continuously in Fuchsia records and is now applied to a form of *magellanica*) being raised by a Mr Bunney by crossing *F. magellanica conica* with *F. microphylla.* This is regarded as improbable. According to M. Felix Porcher, who published in 1848 *Le Fuchsia; son histoire et sa culture,* the first successful hybrid was grown in England in 1837. However in the *Floricultural Cabinet* of 1834, p. 176, W. Barratt, a nurseryman of Wakefield in Yorkshire, gives a list of 26 species and varieties of Fuchsia. Some of these were the result of breeding by Thompson, who had previously worked on the culture of the Pansy. He used principally the different forms of *F. magellanica* available. However he was not the only one active in this field and the same year as Barratt's article appeared, saw the introduction of a cv. to which the name 'Robertsii' was given, in honour of the Cornish gardener John Roberts who raised it. It was of the *magellanica* strain, but had far longer and larger flowers. Barratt's list contains eight recognizable forms of *magellanica,* and species that are named as *lycioides, thymaefolia, excorticata, arborescens* and *microphylla.* The others may be *magellanica* forms or may be hybrids. Presumably the plant called *hybrida* is just that. It is described as a 'very erect growing plant, having pale red flowers'. Whether 'Thompsoniana' is the same plant as that still in cultivation is not easy to determine. W. P. Wood in his *A Fuchsia Survey* (London, Williams and Norgate, 1950) gives 1840 for its introduction; on the other hand the 1835 description sounds quite like our 'Thompsoniana' with its erect habit and rather rosy petals. What Barratt's *F. lucidum* was, even Barratt himself did not know, as he had not yet flowered it. He said that it had leaves like a laurustinus. Similarly unflowered was 'ongiflora' with, according to its anonymous raiser, flowers six inches long. This does sound as though it might contain *F. fulgens,* which had arrived five years previously. The plant is depicted in the 1836 volume and certainly does have an exceptionally long tube, but otherwise it seems to resemble *magellanica* more than anything else. The same plate shows a Fuchsia to which the name of *reflexa* is given: this is one of the small-flowered plants of the Encliandra section.

Although received in 1830, *F. fulgens* does not seem to have been distributed commercially until 1837. The distributor was, again, Lee of Hammersmith. It was immediately used for hybridizing and the modern Fuchsia is basically a cross between *F. fulgens* and the various forms of *F. magellanica*, although, as we shall see, other species have been used.

One of the first of these hybrids was 'Chandlerii', raised by the well-known Camellia enthusiasts. It was a flower of moderate size, with purple, green-tipped sepals and a bluish corolla. It was illustrated in Vol. VII of the *Floricultural Cabinet* with the following interesting note. 'It is a production between *Fuchsia fulgens* and one of the older kinds, probably *globosa.* The seed was from the latter. It is stated by persons who have had ample means of ascertaining, that *F. fulgens* impregnated by the other kinds produces plants with flowers similar to the small kinds, but the smaller kinds impregnated with *fulgens* produce plants having flowers partaking of the form and colour of the latter.' In other words to get large flowers it was advisable to use *F. fulgens* as the pollen parent.

In the same year Colley raised a *fulgens* X *grandiflora* cross. This *grandiflora* was a large-flowered form of *magellanica* and not *F. denticulata* of which *grandiflora* is a synonym. A rather curious hybrid raised by a Mr Wormald (or perhaps Normald) of Ripon, which first flowered in 1839, was a cross between the small-flowered 'reflexa' and *R. aborescens.* It looks singularly un-Fuchsia like with flowers of the *Encliandra* shape but about an inch in diameter, borne in the leaf axils. It was not particularly floriferous and this line of breeding was not followed further. It looks, however, as though a completely new type of Fuchsia could be developed in this way.

About 1839 two new species were introduced into cultivation. *F. cordifolia* was introduced by the London (later the Royal) Horticultural Society. This is an attractive plant with a long tube, green-tipped sepals and a rosy-purple corolla. It does not seem to have been much used in breeding, but 'Exoniensis' and 'Corallina' raised by Lucombe and Pince were both *cordifolia* X *magellanica globosa.* The other species appears under several synonyms and is known now as *F. regia,* but in 1840 as *F. radicans* and also *F. integrifolia.* I do not know that this was employed in hybridization, although given the climate of the period, it would seem improbable that any species would be left untried. We know that the lanky *F. corymbiflora* was used, although we do not know of any cvs. that can be stated to contain *corymbiflora.* Of greater importance to the future breeding of the genus was the appearance of a plant called 'rosea alba', which looks like the well-known *magellanica alba.*

From this Mr Gulliver, the gardener to the Rev. S. Marriott of Horsmonden in Kent, raised 'Venus Victrix'. This has rather small flowers, with a good white tube and sepals and an almost blue corolla. It is still in cultivation and, according to specialists on the genus is the chief ancestor of all the white-tubed and white sepalled cvs. This may well be true, yet the *Floricultural Cabinet* for August 1842 which illustrates this plant, accompanies the illustration with another cv. 'Enchantress' which already looks like a modern white-sepalled hybrid. It is true that there is a pink flush, which is lacking in 'Venus Victrix', but it does suggest that there must have been other white-sepalled cvs. for breeders to work from. In the same journal for November 1841 are depicted four hybrids of *magellanica* cvs. pollinated by *fulgens*. One very strange one has a crimson tube with green sepals and petals and another has a white tube, with white, green-tipped sepals and a rosy corolla. *F. fulgens* itself has green-tipped sepals and some forms have green petals. As these were all first-generation hybrids the white sepals are surprising, as breeders find that the lighter colours tend to be recessive.

After 1840 the number of named cvs. increased rapidly as more and more breeders took the plant up. In 1847, the Frenchman M. Courceles raised the whitest Fuchsia yet, recorded as 'corymbiflora alba'. It is not quite clear whether this is an albino form of *F. corymbiflora* or a hybrid therefrom. In 1843 Veitch received several species from W. Lobb in Peru of which *F. denticulata* (*serratifolia*) has been most used. 1852 saw the introduction by Veitch of 'Dominiana' raised by the famous hybridist Dominy. This was given as *spectabilis* X *serratifolia* (according to modern nomenclature *macrostigma* X *denticulata*). It is still in cultivation at Kew and has long drooping red flowers. It is one of the few existing hybrids without either *magellanica* or *fulgens* in its make-up and it seems a pity that this line has not been pursued.

During the late 1840s and 1850s new species were being received frequently and one appears to have been lost without trace. In 1855 Mr Story brought out four cvs. with white corollas; the two most well-known being called 'Queen Victoria' and 'Mrs Story'. A contemporary account in the *Florist* of 1855 says 'There is no record of the parentage of these varieties; Mr Story had disposed of them to Hendersons of Tollington Nurseries and Mr Story, after a severe illness, has just passed away. It has been stated that he used pollen from a species with an almost white corolla, flowers of which he obtained from Mr Veitch of Exeter, and that the species died after flowering.' Lucombe and Pince of Exeter also brought out some hybrids with white corollas at about the same time, presumably using the same species. It is sobering to think that all the Fuchsias with white corollas are descended from this unknown plant. One does just wonder whether *corymbiflora alba* can have been used. There seems to be no record of its use in breeding, yet it is easily the whitest flower among fuchsias and a good white Fuchsia is still earnestly desired by breeders. It may be sterile, but it looks as though it bears good pollen.

As the century advanced the number of cvs. became immense as more and more nurserymen went into breeding. By this time it would seem as though the actual species were being less used and breeding was between the various hybrid cultivars. Some idea of the number of cultivars being introduced may be gathered by the fact that Lemoine alone introduced over 390 in an eighty-year period; an average of nearly five a year.

There was, however, little in the way of novelty, except in the production of larger flowers: Bull's 'Standard' (1871) had flowers 5 inches long, while its contemporary, Bland's 'Champion of the World', had the largest flowers ever recorded. In the same year James Lye began breeding, using 'Venus Victrix' and its derivatives and produced a large number of cvs. which are still in cultivation today; all with a white tube and sepals of great solidity. In contrast to James Lye, who bred a large number, is Rundle, who produced only three plants, which are still amongst the best. They are called 'Mrs Rundle', 'Mr Rundle' and 'Duchess of Edinburgh'. The first two were said to be derived from selfing one of Laing's cvs. 'Earl of Beaconsfield'.

Towards the end of the century German breeders started working with the recently-introduced *F. triphylla*, which had been found difficult to hybridize. The successful cross was supposed to be *triphylla* X *boliviana*, but some doubt has been thrown on this. The resultant hybrids were in turn crossed with *F. fulgens* and *F. splendens*, the latter similar to *cordifolia* with a scarlet tube and green sepals. These hybrids, mostly bred by Bonstedt, were introduced to commerce between 1902 and 1912. They were dwarf plants and very floriferous, but very frost-tender owing to the *triphylla* blood. 'Eros' is said to have had flowers of a pale salmon-yellow shade and these flowers were not drooping. It seems a pity that it has been lost. In 1929 Hartnauer produced Leverkusen (Leverhulme in Anglo-Saxon countries) which has the excellent *triphylla* habit of perpetual blooming, but is considerably hardier than any other of the *triphylla* crosses. The late Mr C. P. Raffill, when he was at Kew crossed 'Royal Purple' with *F. alpestris* (*F. regia* var. *alpestris*), a semi-climbing Brazilian species, to breed 'Lady Boothby'.

So far as I know, no other species have yet been used in further breeding, although there are a great many available (in theory, at any rate; very few are in cultivation). There is a tetraploid form of *F. magellanica* (var. *macrostemma*) and both *F. coccinea* and *F. lycioides* are tetraploids. *F. triphylla* when examined by Warth in 1925 proved to be a triploid, but it seems unlikely that this can be the general rule. If, however, most of the plants in cultivation come from a triploid clone, it would explain why it was found hard to employ in breeding, as the plant was presumably sterile.

Fuchsia breeding has, since 1929, been practised enthusiastically in the U.S.A. Up to the present I would not say that there has been any marked advance on the European cvs.; certainly nothing so marked as their breeding of Tall Bearded Irises, but since they are near the home of the most valuable species for future breeding, one may hope for a transformation of the flower in the future.

Some of the larger Fuchsias show polyploidy; there are some

octoploids and apparently one has been bred with over 100 chromosomes (presumably a decaploid with 110). Whether this polyploid has any advantage apart from large flowers seems doubtful, owing to the uneven haploid number; however, crosses between hexaploids and diploids might be valuable.

Owing to the large number of unused species there is obviously much scope for future hybridization, though whether this would prove commercially attractive is another matter.

CHAPTER XI

# Viola

The genus Viola is a large one of some four hundred species, which have been divided into seven sections. Only two of these have been developed for horticultural use, although many species are grown as well. The sections developed are *Melanium*, from which the Pansies and Violas are derived, and *Nomimium*, from which the Violets come.

The Pansy (*V.* X *wittrockiana*) appears to be entirely of English breeding originally. In the second decade of the nineteenth century Lady Mary Bennet, Lady Monke and Lord Gambier all started growing the Heartsease, *V. tricolor*. Lord Gambier's gardener was Thompson, who later became a famous plant breeder and it is from the garden at Iver that the true development of the Pansy occurred. The Pansy, it is assumed, is a hybrid between the annual *V. tricolor* and the perennial *V. lutea*. Both these plants are British natives and it is possible, indeed probable, that the two species were not very clearly distinguished. Even in our day subspecies are moved from one to the other and in the early nineteenth century they were all lumped together as the Heartsease. Miller had distinguished them in his *Dictionary*, where *V. lutea* appears as *V. calcarata*. He had also commented on the variability in flower size of *V. tricolor*.

In 1841 the *Floricultural Cabinet* extracted an article by Thompson from the *Flower Gardeners' Library*, which details the early history of the Pansy. Since this was originally almost entirely Thompson's creation, it seems worth while reproducing most of this piece.

'About twenty-seven or twenty-eight years ago, Lord Gambier brought me a few roots of the common yellow and white

Heartsease, which he had gathered in the grounds at Iver, and requested that I would cultivate them. . . . I did so, saved the seed, and found that they improved far beyond my most sanguine expectation. In consequence thereof I collected all the varieties that could be obtained. From Brown, of Slough, I had the blue; and from some other person, whose name I do not now recollect, a darker sort, said then to have been imported from Russia. These additions wonderfully improved my breeders. But still, though the varieties I soon obtained were multitudinous, their size was almost as diminutive as the originals. . . . Up to this period, which was about four years [i.e. *c* 1816] after my commencement, I began imperceptibly to grow pleased with the pursuit, for all who saw my collection declared themselves delighted therewith. I then began to think that some of my sorts were worthy of propagation; and this circumstance led me to give one, which took his lordship's fancy a name. This was entitled Lady Gambier, and as I struck cuttings of it, they were given as presents by my worthy employers to their numerous friends and acquaintances. . . . This, though in comparison with the worst flower now grown (and many, even of the named varieties, are bad enough), would even beside them be reckoned little better than a weed. Still Lady Gambier was the beauty of her tribe, and won golden opinions from every beholder. It was, indeed, in shape little more symmetrical than a child's windmill, but looked in size among the sisterhood like a giant surrounded by dwarfs. But the giant of those days would be a pigmy now, as Lady Gambier herself appeared in comparison with another flower, which I soon after raised, and which, on account of what I then considered its monstrous proportions, I christened Ajax. This I then thought could never be surpassed, and yet in shape it was as lengthy as a horse's head.

'Still I had worked wonders, and I resolved to persevere. I did so and was at length rewarded by producing rich colouring, large size, and fine shape. The first large and good shaped flower that I raised was named Thompson's King. Still, up to this period, a dark eye, which is now considered one of the chief requisites in a first-rate flower, had never been seen. Indeed such a feature had never entered my imagination – nor can I take any merit to myself for originating this peculiar property – for it was entirely the offspring of chance. In looking one moment over a collection of heaths, which had been some time neglected, I was struck, to use a vulgar expression, all of a heap, by seeing what appeared to me a miniature impression of a cat's face steadfastly gazing at me. It was the flower of a Heartsease, self-sown, and hitherto left to "waste its beauty far from mortal's eye". I immediately took it up and gave it "a local habitation and a name". This first child of the tribe I called Madora, and from her bosom came the seed, which after various generations produced Victoria, who in her turn has become the mother of many even more beautiful than herself. Hitherto, in the way of colour, nothing new had been introduced; white, yellow and blue in their numerous shades, seemed to be the only colours which the Heartsease was capable of throwing out, till about 4 years since, when I discovered in my seedling bed a dark bronze flower, which I immediately marked and baptized Flamium; – from this have sprung Tartan, Vivid, and the King of Beauties, which has only bloomed this spring, and is, decidedly, the best flower of its kind that has ever been submitted to public inspection.'

The 'darker sort, said to be imported from Russia', was probably *V. altaica* var. *purpurea,* which was introduced in 1810. It would seem to follow therefore that the Pansy is a hybrid of three species. Its primary development would seem to show hybrid vigour followed by gigantism. There seems no reason to believe that any of the Pansies are tetraploid and gigantism is a fairly common phenomenon with cultivated plants. It is interesting to compare the illustrations of Pansies in the early 1830s with those depicted at the end of the same decade. The early ones still show the elongated shape of *V. tricolor* and *lutea* and, to my eyes at any rate, are more attractive than the circular blooms that were developed and which have been preserved to today. Incidentally, in 1839 someone was advertising a Scarlet Heartsease, but nothing more is heard of this and it was probably an example of questionable sales promotion.

By the 1840s the Pansy had become Show or Florists' Flowers. They were classed in two categories with strict rules. There were the 'Selfs' in which all the petals had to be of the same colour, although the 'eye' was allowed to be of a different colour, so long as it was not too conspicuous. The other category was the White or Yellow Ground. These were essentially bicolors, but the two colours had to be strictly controlled. The three lower petals had the ground colour, whether white or yellow, but they were bordered by another colour, which also had to suffuse completely the two upper petals; the 'eye' was also supposed to be of this second colour. Such flowers were very attractive in themselves, but the amount of variation permitted was so limited that it is not surprising to find someone writing in the *Gardeners' Chronicle* on April 5, 1862, 'The Show Pansies seem to have nearly run their length, owing to the want of variety among them, for though the so-called novelties are still introduced in considerable numbers, . . . yet, as far as the public can appreciate, they present year after year little more than repetitions of stereotyped forms. . . . We have not infrequently seen stands of 24 blooms, in which 8 or 10 were virtually alike.'

In the 1830s the Pansy was introduced to France and later to Belgium and the growers there, not hamstrung by the artificial rules of the show bench, bred plants for size and colour with no regard to regularity. These so-called Fancy Pansies were brought to England by John Salter, who had a nursery at Versailles as well as one at Hammersmith. He also had the courage to persist in the face of public indifference with a plant he thought was good. He had championed the ragged Japanese Chrysanthemums at a time when the public thought such raggedness barbarous and he persisted in the Fancy Pansies for ten years before they aroused any enthusiasm. Eventually they seem to have surpassed the Show Pansies altogether, although these still linger on in the hands of a few enthusiasts. However, in general, the Pansy is now regarded as a bedding, rather than as a show flower, and treated as a hardy biennial. There still exists the North of England Pansy and Viola Society, the members of which compete among themselves with show blooms.

In his historical chapter in *Pansies, Violas and Violets* (London, 1898) William Cuthbertson quotes from letters written to him by James Grieve, when he was breeding for Messrs Dickson and Co. In 1862 he started by crossing Show Pansies with *Viola lutea*; in 1863 he crossed Pansies with *V. amoena* and *V. cornuta*. Next he procured *V. stricta* and thereby raised a number of flowers without rays or blotches. In 1867 the nursery obtained plants of an improved form of *V. cornuta* called 'Perfection' and Mr Grieve 'crossed every bloom with everything he could lay his hands on' and had 700 seedlings as a result. *V. amoena* seems to be a name for the purple form of *V. lutea. V. stricta,* also known as *V. hornemanniana,* is a puzzle. Loudon says that it has cream flowers, Sweet that the flowers are pink or blue; its country of origin seems unknown, and it was introduced either in 1819 or 1822. It would seem from this account that the usual explanation that the Viola was the result of crossing the Pansy with *V. cornuta* is not strictly accurate, although, as we shall see, it is certainly correct for some strains.

*Viola cornuta* is a Pyrenaean plant of tufted growth and medium-size purple-violet flowers with, as the name implies, a long spur. White forms are also known. This is certainly the principal species used in breeding the Viola (which William Robinson in vain tried to christen the Tufted Pansy) which is distinguished from the Pansy by its smaller flower, often without a central blotch and eventually without even central rays, and its very long flowering season. If prevented from seeding, it will continue flowering throughout the summer and autumn. The raiser of the first rayless Viola was Dr Charles Stuart, who lived at Chirnside, Berwick, and fortunately we have his account of its raising.*

'In 1874 I took pollen from a garden pansy named Blue King . . . and applied it to the pistil of *Viola cornuta.* There was

* In *Pansies, violas and violets,* by W. Cuthbertson (London, 1898).

a podful of seed which produced twelve plants which were well taken care of. The next season they flowered and were all blue in colour, but with a good tufted habit. I again took the pollen from a pink garden pansy and fertilized the flower of my first cross with a limited success. The seed from this cross gave me more variety in colour of flower, and the same tufted habit of growth, which evidently came from the *V. cornuta* influence. The best of this cross were propagated and grown, some of the plants being sent to the R.H.S. gardens at Chiswick for trial. . . . After being in the ground for some time I received a letter from a member of the Floral Committee, inquiring how they had been raised, as they were entirely different in growth from all the others sent in. . . . I was rather surprised when informed that I had got six first-class certificates and was first in the competition. Nothing more was done at this time, beyond growing the plants I had already raised, and sowing the seed from them in a bed broadcast. They were all more or less rayed. A floral ally, seeing one of these certificated plants, a fine white Self, remarked: "If you could only get that flower without rays in the centre, I think it would be a great improvement." Keeping a sharp look-out on the seed beds, it was ten years before I succeeded in finding a really rayless Viola. In the year of the Queen's Jubilee, while walking round the seed-bed, I saw what I had been seeking for, in a pure white rayless Self. The plant was there and then pulled to pieces and every bit propagated. . . . Such is the true history of "Violetta", one of the most popular of the rayless tufted Pansy family. "Violetta" has proved the mother of thousands of a rayless race, some better, some worse than the parent.' Dr Stuart goes on to explain how he crossed one of these rayless Selfs with a Fancy Pansy, but adds 'I found out, however, that this Pansy crossing was too much, for out of a hundred and fifty seedlings "Border Witch" was the only one without rays.'

Another observation of Dr Stuart's is worth quoting. 'In hybridizing or crossing wild varieties of Violas, it is necessary that the pollen be taken from the cultivated species of Pansy and dusted over the pistil: that is that the wild species should be the mother. Pollen taken from *V. cornuta,* for instance, will, if put on the common Garden Pansy, only give seed which will produce Bedding Pansies, not the sturdy tufted rooted dwarf strain.'

Another breeder who left some notes on his work was Richard Dean, who comments among other things that the early Violas were not successful in the south of England and the first that was successful in this region was 'Blue Bell', a chance seedling of his own. He worked by crossing Pansies with *V. cornuta* 'Perfection', a clone sold by B. S. Williams, *V. lutea* 'grandiflora', which was presumably another selected clone and the Cliveden Purple Pansy, whatever that was.

It would seem, then, that the Viola was mainly the Pansy (*V.* X *wittrockiana*) crossed with *V. cornuta,* but that it was occasionally back-crossed on to *V. lutea* and some strains also contained *V. stricta.* According to James Grieve, *V. stricta* gave rise to plants without rays.

Of these species *V. cornuta* is a diploid with 22 chromosomes, *V. tricolor* a diploid with 26 chromosomes, while *V. lutea* varies between 26 and 53 chromosomes. The genus *Viola* has four basic haploid counts of 6, 10, 11 and 13, while some species have 6 plus 11, while others are tribasic with 6 plus 10 plus 11. This seems to result in the creation of fertile hybrids between species which would appear to be incompatible. One would expect the pollen of the Pansy to contain 13 chromosomes, while the ovule of *V. cornuta* would only contain 11, but the resultant hybrid is, as we have seen, fertile.

The development of the Pansy and the Viola is well documented; we find a completely opposite state of affairs when we come to consider the Violet. These plants have been cultivated over a very long period, particularly in Persian and Turkish gardens. The species used would probably not have been our native *V. odorata,* but the related *V. alba,* which replaces *V. odorata* in Eastern Europe. The late E. A. Bunyard writing in the *New Flora and Silva** mentions the Arabo-

* Vol. 4, pp. 187-94 (1932).

Spanish writer Ibn el Awam as contrasting the large-leaved cultivated Violet with the small leaved 'mountain' Violet and concludes that large flowered forms may already have been developed at that time. By the late sixteenth century the herbalists were referring to double forms of *V. odorata* and to various coloured forms, white, pink, blue and yellow. These seem to have been cultivated with little variation for two hundred years.

In the early nineteenth century *V. suavis* was introduced from Russia (syn. *V. sepincola, V. pontica*); it is a plant with rather pale blue fragrant flowers, which are larger than those of *V. odorata.* It is not clear whether the 'Russian Violets' referred to in contemporary literature refers to forms of this plant or to hybrids between it and *V. odorata.* Another form of *V. suavis* was brought from Turkey about 1860 by M. Levallée and was grown on the Riviera and at Oran under the cultivar name 'Wilson'. In France the Violet was associated with Bonapartism and it was cultivated for political reasons as much as for its decorative qualities. In 1835 a M. Jean Chevillon found the plant known as var. *praecox* of *V. odorata,* which flowers intermittently from October to April. This was known as the Violette de Quatre Saisons and was bred into the normal strain. Meanwhile in England F. J. Graham of Crauford had raised a Violet with exceptionally large flowers. This was 'Czar', which was given an Award of Merit by the R.H.S. in 1865. The name suggests that it was bred from a Russian Violet. Nothing like this had been seen before and it became extremely valuable, not only for its own sake, but because it transmitted its great size to its progeny. 'Czar' crossed with the old favourite 'Devoniensis' gave 'Victoria Regina'. Although by this time most of the breeding was done in France, mention should be made of 'Wellsiana', introduced in 1886, which had flowers an inch in diameter.

Meanwhile in France A. Millet, the historian of the Violet,* and his father had been breeding extensively at their nursery at Bourg-la-Reine, near Paris. The form of *V. suavis* known as

* *Les Violettes* (Paris, 1898).

'Wilson' was crossed with 'Czar' to give 'Luxonne', which was intensively grown for scent around Hyères in the late nineteenth century. Millet had introduced the lilac-coloured 'Lilas', and this crossed with 'Czar' gave 'Brune de Bourg-la-Reine' which was of an odd metallic tinge. 'Princesse de Galles' was a very large Violet; it was, however, overtopped by 'La France' which had flowers as large as a Viola. Towards the end of the century breeders had the idea of using the North American *V. papilionacea,* which has large well-shaped flowers, unfortunately devoid of perfume. It is presumably to this species that must be attributed some large-flowered cultivars that are scentless. In his article* Mr Bunyard suggests that the large-flowered Violet was the result of gigantism and was therefore not dissimilar to the nearly contemporary appearance of the giant Cyclamen.

A particular problem is posed by the so-called Parma Violets. These are double-flowered, of a lavender-blue colour and have a different perfume to the ordinary *V. odorata.* In the early nineteenth century they were known as Neapolitan Violets and there was a tradition that they came originally from Turkey. Millet says that they were for many years to be found only in Royal Gardens, but omits to state which Royal gardens; to add to the confusion the Neapolitans call them Portuguese Violets. The authors of the introduction to *Viola* in the R.H.S. *Dictionary,* suggest that *V. alba* is the probable ancestor of this Violet. It has been in cultivation for at least two hundred years and probably much longer, and being sterile has been propagated entirely by vegetative propagation. The English stocks recently showed signs of a virus infection and it is doubtful if it still exists in this country. It has probably been maintained on the Continent. Although other double Violets are known, they differ in colour and perfume from the Parma Violet and if this is lost it can probably not be recovered. This is always a danger with double sterile flowers and it is indeed remarkable that the plant should have survived for so long. In Edwardian days they were very popular

* *In New Flora and Silva,* vol. 4 (1932).

but nowadays their popularity has waned and, since they are not very floriferous compared with other cultivars, stocks have been allowed to diminish. It would seem in any case that the modern cultivars are inferior to those grown at the end of the nineteenth century and there has presumably been a lack of commercial demand for them.

It would appear, in fact, that all the races of *Viola* are falling into disrepute. Pansies are no longer sold with cultivar names and, although Violas still are, they are not being bred with any enthusiasm. There are few growers of Violets and it would be interesting to learn when the last cultivar was introduced. The Show Pansy must have nearly the shortest history of any florists' flower.

CHAPTER XII

# The Begonia

The development of the tuberous Begonia can be ascribed almost entirely to the firm of James Veitch & Sons. With a couple of possible exceptions they have all been raised from six species, five of which were introduced by Veitch and three of which were found by Richard Pearce, one of their collectors in South America. His first introduction was *Begonia boliviensis,* which he collected in Bolivia in 1864 and which was shown in public in 1867. This is a tall growing plant, with annual stems reaching up to 2 feet and rather small drooping flowers of bright scarlet. It is not one of the most attractive species, but its influence may be discerned in the pendulous Begonias that are cultivated for growing in hanging baskets.

Pearce's second introduction, which was named in his honour *B. pearcei,* was also his most important. This is characterized by clear yellow flowers and handsomely marbled leaves and it is the principal parent of all the yellow-flowered Begonias. It was also a native of Bolivia and was sent home in 1865.

Pearce's next introduction, *B. veitchii,* was of nearly equal importance, as the flowers are well-rounded in shape and of good size. The flowers are scarlet and the leaves peltate. Pearce sent this back in 1867 and it was put into commerce in 1869.

1867 had seen the flowering of two other species which were not introduced by Pearce. *B. clarkei* had been purchased from E. G. Henderson & Son as an unknown Peruvian plant by Colonel Trevor Clarke and it flowered with him for the first time in 1867. The plant is not very dissimilar to *B. veitchii* in appearance, but is rather taller and is tenderer, needing more heat to bring it to perfection. *B. rosaeflora,* a Veitch importation from the Peruvian Andes, has good round-shaped flowers of a deep pink colour.

The last of Veitch's importations was considerably later in its introduction. Collected by Mr Davis, and named in his honour *B. davisii,* it flowered for the first time in 1876 and was not introduced into commerce until three years later. This is a delightful dwarf species, which does not throw up tall stems like the other tuberous species, but bears its medium-size bright scarlet flowers on long peduncules, well overtopping the foliage. It is the principal parent of all the 'bedding' Begonias, as its compact habit is transferred to its offspring.

These six species all hybridized well together and the hybrids could be further crossed amongst themselves. A seventh species *B. froebelii* which flowers in autumn and winter was introduced in 1872 and much was hoped from it. However it proved incompatible with the other six species and only two hybrids are recorded, as far as I know; with *B. dregei* and with *B. polypetala.* I can find out nothing about this last species which seems no longer to be in cultivation.

There is yet one more species involved in the hybridizing of tuberous Begonias, but no one knows what it is. The first hybrid put into commerce, raised by Veitch's foreman Mr Seden, was *B.* X *sedeni,* which was sent out in 1870. This is described as a cross between *B. boliviensis* and an Andean species 'Which the Messrs Veitch then had, but which was never named nor sent out' ('The Tuberous Begonia,' *Gardening World,* 1888).

As a general rule it may be said that the tuberous Begonias are not easy to cross with the non-tuberous varieties, but one of the earliest white Begonias was 'White Queen' which was the result of crossing the shrubby *B. parvifolia* with *B.* X 'sedeni'. This was raised by Mr James O'Brien. He tried a similar cross of *B. parvifolia* with *B. veitchii,* but though he succeeded in raising plants (described as having *veitchii* tubers and shrubby stems) the buds invariably dropped before opening.

O'Brien was also responsible for raising the first two double Begonias in Great Britain, but the plants were subsequently lost. Indeed the aim of Seden and the other early hybridizers does not seem to have been directed towards the double flowers that are now *de rigueur* with florists' Begonias, but towards variation in colour and increased floriferousness. Another of Seden's hybrids, was 'Excelsior,' which included *B. cinnabarina,* a species no longer in cultivation (there is a plant sometimes sent out under this name, but it is a cultivar of *B. fuchsioides*). One of Seden's most important hybrids was 'Queen of Whites' distributed in 1878 and derived mainly from *B. rosaeflora.* The official description is from 'light-coloured varieties' of this species, but it is doubtful if variety is used in the sense that we use it today. Probably it was the offspring of two *rosaeflora* hybrids that were very pale.

It was not long before continental raisers were also engaged on the Begonia and about 1875 they began to turn their attention to raising double Begonias, which had not hitherto been much esteemed. We soon find the expected names in the introducers of these new flowers: Lemoine, Felix Crousse and Louis van Houtte. Although Lemoine is particularly associated with Begonia hybrids, his most successful work lay with the shrubby and rhizomatous types and it was Felix Crousse who developed the large double tuberous Begonia. Even so the first outstanding plant in this group was Lemoine's 'Lemoinei'. In this country John Laing, a nurseryman at Forest Hill, Catford, started breeding in 1875 with the best British and Continental hybrids he could obtain, and as a result of his work, British-raised Begonias became equal to the best continental ones.

In 1892 a new species, originally found in 1866, was introduced into commerce. This was *B. baumannii.* Seeds of this were sent by its discoverer, Dr Sace, to the Alsatian nurserymen Baumann, who sold their stock to Lemoine, and they distributed it in 1892. This is an attractive tuberous Begonia, somewhat like *B. rosaeflora,* but with the advantage of possessing a strong perfume. In cultivation this did not seem to be a constant feature, some plants being strongly scented, some faintly and others completely odourless. Nevertheless breeders, particularly G. A. Farini (author of *How to Grow Begonias*, London, 1896), attempted to breed in this species in order to produce a scented Begonia. If they were successful, their products seem to have disappeared and I know of no scented hybrid Begonia, which is not surprising in view of the comparative rarity of the fragrance of *B. baumannii* in cultivation. It would quite possibly be worth a breeder's while to re-introduce the species and try again.

Begonias are monoecious and it is only the male flower that is double: the stamens are converted into petals. Some of these doubles, if starved, can be coaxed into producing a few stamens and can be used for breeding; others remain completely sterile, whatever one does, and so are useless for further breeding. However if all the seedlings from a cross are kept, even the singles will contain all the genes of the original cross and may be used both for a $F^2$ generation and for breeding with other hybrids. Apart from increased size little seems to have been accomplished since the turn of the century.

In 1880 Sir Isaac Bayley Balfour introduced *B. socotrana*. This is a native of the Island of Socotra in the Indian Ocean. Although present in many hybrids, the species is very rare in cultivation. The stems arise from scaly bulbils, which are formed around the base of the previous year's stems. These stems reach a height of 6-9 inches and carry dark green peltate leaves and terminal inflorescences of bright pink flowers that are up to 2 inches in diameter. Owing to the climate of Socotra, with its very hot dry summer, the plant starts into growth in September and flowers during the winter. All breeders were aware that this winter-flowering character was very valuable and they naturally tried to hybridize it with various species. Since the tuberous Begonias were the most popular these were the first tried, although the affinity with the South African Begonias of the section *Augustia* was at once evident. The first hybrid, named 'John Heale', after its raiser, a foreman in Veitch's nursery, appeared in 1885. It was the result of a cross between *B. socotrana* and 'Viscountess Doneraile'. 'Viscountess Doneraile' has an involved parentage being derived from

'Sedeni' and 'Monarch'. 'Monarch' came from 'Sedeni' crossed with 'Intermedia' and 'Intermedia' was *B. boliviensis* X *B. veitchii.* In fact, then, 'Viscountess Doneraile' is a secondary hybrid of *Bb. boliviensis, veitchii* and the unknown second parent of 'Sedeni'. 'John Heale' was a very dwarf plant with cordate, acuminate leaves and carmine flowers, which, somewhat oddly, are all male. It came into flower about mid-October, which was rather too early. Two years later Heale produced two further crosses: 'Adonis', the result of pollinating some unnamed tuberous hybrid with the pollen of 'John Heale', and 'Winter Gem' which was a *socotrana* cross with, again, an anonymous scarlet tuberous Begonia. 'Adonis' had flowers twice as large as 'John Heale', but these were also entirely male. 'Winter Gem' had scarlet flowers, produced bulbils like its parent and had flowers of both sexes. Later crosses with tuberous Begonias were made by other British and Continental breeders and a few survive in cultivation, but owing to the smaller number of greenhouse owners they are not very popular. They flower during November and early December and are over by Christmas, so that they have little commercial value; and since they are also very prone to get Begonia Mildew, they are not attractive to amateurs. Very different is Lemoine's cross of 1892 'Gloire de Lorraine', the result of crossing *B. socotrana* with the South African *B. dregei.* This latter species has annual stems that may exceed 2 feet in height and white flowers in large panicles. The hybrid has taller stems than *B. socotrana* itself, but has the colour of *socotrana* with the floriferousness of *dregei* and, since it is easy to have the plant in perfect condition at Christmas, it must be regarded as one of the most successful hybrids ever raised. There are other Lemoine *socotrana* hybrids, 'Gloire de Sceaux' (*socotrana* X 'subpeltata') and two plants of the *socotrana* X *roezlii* cross: 'Triomphe de Nancy' and 'Triomphe de Lorraine'. These are useful plants but have not obtained the popularity of 'Gloire de Lorraine'. There are some colour variants of the latter plant with white and carmine flowers and with dark purple leaves, but the original Lemoine hybrid has remained the most popular.

A curious modern use of *B. socotrana* may be mentioned in the cultivar 'It'. This is a second generation cross of *B. socotrana* with a Rex Begonia, not presumably the original *B. rex*, but one of the hybrids derived from this and *B. diadema*. 'It' has ornamental leaves, although not so ornamental as the best of the Rex hybrids and reddish flowers that are a distinct improvement on the dirty white flowers of the Rex group.

Veitch and Lemoine did a lot of interspecific crossing of rhizomatous and shrubby Begonias and raised many valuable plants, but they are not sufficiently removed from their parents to warrant investigation here. On the other hand the crossing of *B. rex* and *B. diadema* has resulted in the creation of a series of extremely handsome foliage plants, commonly known as Rex Begonias. Besides the nurserymen mentioned, there were many Germans who were interested in raising cultivars and the Belgians were also engaged, so that the race may almost be regarded as Pan-European. *B. rex* has leaves with a smooth edge, whereas the leaves of *B. diadema* are incised. It seems likely that the red-leaved *B. cathayana* may have been bred into this strain and certain that *B. decora* with reddish-brown leaves was. Both *B. cathayana* and *B. diadema* are fibrous-rooted and shrubby, while *BB. Rex* and *decora* are rhizomatous and do not throw up tall stems; the leaves spring directly from the rhizome. The cultivars with *cathayana* and *diadema* blood make taller plants than the *rex* X *decora* plants and *diadema* also gives the hybrids jagged leaves. *Begonia rex* was introduced in 1858, but the other species much later; *diadema* in 1883, *decora* in 1896 and *cathayana* in 1902. The race is therefore comparatively modern and, as in the case of the tuberous Begonias, there has been little progress made after the first flush of cultivars.

One disadvantage of intensive hybridization is that the original species get lost. Any breeder wishing to reintroduce any of the original parents would find it very difficult to obtain living plants. The R.H.S. *Dictionary of Gardening* says, after its description of *B. veitchii* 'should be reintroduced'. One would expect Botantic Gardens to maintain these parental species, but greenhouse space is limited and they do not always do so. *Begonia cathayana* is still grown, but I have yet to see pure plants of the other parents of the Rex group. It would seem probable that the best chance of further progress in Begonia hybridizing would be by bringing in fresh genes from newly-collected species. Fresh plants of *B. pearcei* might revolutionize the yellow Begonias. *B. cinnabarina* appears to have been lost before its full potentialities had been exploited. It is by no means certain that reintroducing the original species would have effective results, but it seems the most promising line of progress. It is true that the plant is no longer as popular as it was, but this may be partly owing to the lack of progress. The most popular of the Begonias nowadays are the cultivars of *B. semperflorens* and there may be room for improvement here. They have been crossed with tuberous Begonias to give a small-flowered tuberous bedding plant, but, being fibrous rooted, the potentialities for further hybridization seem to be considerable.

Very little work appears to have been done on the cytology of *Begonia* and the hybridist is rather on his own. However this may not mean too much. *B. socotrana* has a count of 28, while *B. dregei* has a count of 26, so in theory 'Gloire de Lorraine' is a very unlikely hybrid. What appears to happen is that *socotrana* belongs to a group with a haploid count of 7 and is a tetraploid, while *dregei* belongs to a group with a haploid number of 13, which divides into 6 plus 7. Possibly in the cross the 7 chromosomes of *dregei* combine with the 14 of *socotrana* and the odd 6 drop out. Whether the chromosome count of 'Gloire de Lorraine' is 42 does not seem to have been established, and my explanation may be quite incorrect. Begonias seem to have a series of haploid numbers of 6, 7, 9, as well as the group of 6 plus 7. It is curious that of all the species whose counts have been published there is not a single diploid except in the 6 plus 7 group. *Begonia gracilis*, a little-known tuberous species from Mexico is a dodecuploid with a count of 84. These high counts suggest a large number of different genes, but it would obviously be necessary to raise very large quantities to bring out all the recessives. With a few exceptions Begonias do not show excessive variation among individual species and a high chromosome count does not necessarily mean that the plant is variable. The common Stinging Nettle has $2n=52$ and is a very stable species. There is no essential correlation. Indeed variation seems to be more associated with a high haploid number (as e.g. *Dactylorhiza* with 20), than with polyploidy. However that may be, there is some variability in Begonias in respect of colour both in the flowers and in the leaves and the ease with which certain species hybridize suggest that much remains to be done. It is a pity that the cytologist cannot help the breeder more with this genus.

CHAPTER XIII

# The disappearing species

*Aquilegia – Astilbe – Bouvardia – Calceolaria – Canna – Clematis – Dahlia – Delphinium – Freesia – Gladiolus – Hemerocallis – Hippeastrum – Lupin – Mimulus – Montbretia – Nymphaea – Pentstemon – Petunia – Phlox – Schizanthus – Streptocarpus – Verbena. Deciduous Azaleas – Erica*

During the nineteenth century not only were new species being brought into the country in increasing numbers, but nurserymen had begun to hybridize everything that they had growing in their nurseries. These hybrids seemed in many cases preferable to the original species. They were more vigorous, or they had a greater variety of colour or seemed in other respects more desirable than the species from which they were bred. As a result the original species dropped out of cultivation and only the hybrid races remained. Nowadays there are a number of hybrid herbaceous plants common in gardens, while their ancestral species are either no longer in cultivation or extremely hard to come by. A typical example is the Columbine.

### ❁ *Aquilegia*

Since it is extremely difficult to preserve the species of *Aquilegia* unhybridized if more than one is grown in the garden, it is not surprising that the popular garden Columbines are generally of hybrid nature. The long-spurred hybrids are thought to be the result of crosses between three North American species *AA. chrysantha, canadensis* and *formosa.* Both *chrysantha* and *canadensis* are mainly yellow flowered, although the spurs of *canadensis* are tinged with red. *A. formosa* has a scarlet and yellow form as well as a plain yellow and it is this form which has been chiefly used in the breeding. The old English Columbine is *A. vulgaris,* usually blue in the wild state, but showing a large amount of variation over its extensive range. The flower has rather short spurs and crossed with the long-spurred American species has given rise to what are termed 'short-spurred' hybrids, but the spurs are considerably longer than in the wild type. This has also brought blue and pink shades into the strain. Since *A. vulgaris* is a more vigorous plant than the others, it will often be found that, if the short-spurred hybrids are allowed to seed, eventually the offspring will resemble *A. vulgaris* altogether. In order to preserve the best forms, it is necessary to rogue the parent plants continually. The Clematis-flowered hybrids are spurless with very open flowers resembling a clematis. They may just be a selected form of *A. vulgaris* or they may be a hybrid between *A. vulgaris* and the Himalayan *A. fragrans,* which has a very open flower and is short-spurred.

Bailey in his *Hortus** suggests that the garden hybrids contain, in addition to the species already mentioned, *AA. glandulosa, sibirica, caerulea* and *skinneri.* Of these *glandulosa* and *sibirica* are low-growing plants, while *skinneri* is a Mexican species and tender in Great Britain. It might well have entered into American-bred hybrids and the same can be said of the very lovely *A. caerulea,* which is not easy to grow in this country, but which would certainly be desirable for introducing blue colours and long spurs into the garden race. I would be inclined to question whether *glandulosa* and *sibirica* had been deliberately bred into any strain, but they could well have entered unintentionally. Unless seed is collected in wild locations, it is often very difficult to obtain pure species nowadays. Any nursery growing the hybrid strain and any other species would probably find this species hybridizing with the hybrids, with the result that modern plants may contain more species than the breeder intended.

* L. H. and E. Z. Bailey, *Hortus* (New York, 1930).

### ❀ *Astilbe*

The popular Spiraeas of gardens and forced plants are principally of hybrid origin. The cross-breeding seems chiefly to have developed a rather larger colour range, as the hybrids themselves do not differ in external appearance from the species. The hybrids are in two main races, although others are recorded. The most spectacular are the Arendsii hybrids, raised by the German nurseryman Arends around 1907. These all have as one parent the very tall *A. davidii,* which can reach 6 feet and has rather fierce magenta flowers. Crossed with white or pink species, the resultant plants were in delicate shades of pink or red and of more reasonable dimensions. One of the largest of these hybrids is *A. davidii* X *A. thunbergii.* This latter plant is only about 18 inches high so it is rather curious that in this cross alone the tall habit of *A. davidii* has been perpetuated. The recorded crosses are *A. astilboides* X *A. davidii,* with rosy lilac flowers; *A. davidii* X *A. japonica* with white or very pale pink flowers; *A. davidii X A. X rosea* with a range of colours ranging from purple to white and the cross already mentioned *davidii* X *thunbergii,* reaching a height of 4-5 feet with white or pale pink flowers.

The other main race is *A.* X *rosea,* which is *A. chinenis X A. japonica.* From this come such cultivars as 'Peach Blossom' and 'Queen Alexandra' in various shades of pink. *A. chinensis* is the larger species with pinkish flowers on a fairly tall spike. *A. japonica* is dwarfer and rather tender; it has pure white flowers. Another hybrid is *A.* X *lemoinei* which is *A. astilboides* X *A. thunbergii. A. astilboides* is a dense-flowered white plant and since *A. thunbergii* is also white, it is surprising that the hybrid has a pinkish tinge. Perhaps the parentage was not correctly recorded. Another Lemoine cross, 'Rubella', had *davidii* as one parent and seems similar to Arends's *davidii X X rosea* cross. At one time there were a series of hybrids under the name of *A.* X *crispa,* which were dwarf plants with *A. chinensis* var. *pumila* as one parent and perhaps the dwarf *A. simplicifolia* as the other. There were also crosses between this last-named species with the *rosea* hybrids. These dwarf forms may no longer exist, as few modern gardeners wish to grow hybrids in the alpine garden. They appear to have been mainly of German breeding.

### ❀ *Bouvardia*

This genus, a member of the *Rubiaceae,* was at one time grown in great quantities as its panicles of flowers were very freely produced, long-lasting and were much used for decoration. They require warm greenhouse treatment and, as a result, are now far less grown than they were when fuel and labour were cheap. The first hybrids were made in this country in 1857 and were the result of crossing *Bouvardia longiflora* with *B. leiantha. B. leiantha* has corymbs of scarlet flowers, which are rather small, while *B. longiflora* has large, white fragrant flowers, which are produced in few-flowered corymbs. The resultant hybrids (two of which were christened 'Hogarth' and 'Laura') had flowers intermediate in size in shades of pale pink and salmon. Subsequently other species were crossed with these hybrids, but apart from the influence of the few-flowered *B. flava,* which introduced a yellow colour into the hybrids, there would not appear to have been much improvement of the original cross. An exception is the plant with glossy green leaves and long tubular fragrant flowers, which has been given the name *B. humboldtii corymbiflora.* This looks like a hybrid between the two white fragrant species *B. longiflora* and *B. jasminiflora.* It has the pointed leaves and winter-flowering habit of the last-named, with the large flowers of *B. longiflora.* At some period 'Hogarth' must have produced a pale-pink sport, which was propagated under the name 'Bridesmaid'. This must have been what it called an autogenous chimaera, as it affected only the exterior cells. Bouvardias can be propagated by root cuttings and the buds from root cuttings come from the internal tissues of the plant. Thus root cuttings of 'Bridesmaid' always produce 'Hogarth', showing that it is only the external parts of the plant which have sported; the interior tissues have remained unchanged.

Although the original cross was made in England, later hybrids were made in all countries and most notably, in the U.S.A. The two most popular cultivars today are 'President Cleveland' and 'President Garfield'.

### ❀ *Calceolaria*

The silence of all reference books on the subject of the breeding of the hybrid Calceolarias would suggest that it is regarded as a shameful secret to which no reference should be made. So far my sole source has been an article in the *Floricultural Cabinet* for April 1841. According to this writer, the first hybridizer was Mr Penny, foreman at the nursery of Messrs Young of Epsom. His first hybrid, 'Gellaniana', was *C. corymbosa* X *C. purpurea*; and his second, 'Youngii', was *C. corymbosa* X *C. arachnoidea.* Subsequent to the introduction of *C. crenatiflora* in 1831, this too was bred into the strain. These are all herbaceous forms and it would seem probable that the modern herbaceous hybrids are principally raised from these species.

*C. corymbosa* is a slender perennial about 18 inches high with yellow flowers about half inch across. The flowers have faint purple stripes. *C. crenatiflora* is similar, but a more robust plant with larger flowers. *C. arachnoidea* is a low plant with the leaves and stems covered with a cobweb-like indumentum. The flowers are small and purple, and there are not many to a panicle. *C. purpurea* is a vigorous plant up to 2½ feet in height with dull purple flowers, about ½ inch across, in a much branched panicle.

It would not seem that *C. arachnoidea* can have been much used in the original crosses as there seems no record of any cobwebby leaves. In modern times another species with cobwebby leaves has been bred into the strain. This is *C. cana,* a moderately tall plant with small but fragrant flowers. These are either purple or yellow, depending on whether you read pp. 354 or 355 of the first volume of the R.H.S. *Dictionary of Gardening.*

The shrubby varieties, that are still used for bedding, although less frequently than in Victorian times, are based on *C. integrifolia*. This can, apparently, reach a height of 4 feet and has sticky leaves and panicles of small yellow flowers, in the typical form. Cultivars exist with brown or orange flowers and it is not clear whether they are selected colour forms or hybrids. Another species used in Victorian bedding schemes was *C. amplexicaulis*, a more compact plant with slightly larger flowers than *C. integrifolia*. It seems, therefore, probable that the bedding Calceolaria is basically a hybrid of these two species. It would surely have been impossible for Victorian nurserymen to resist hybridizing the two species.

The article in the *Floricultural Cabinet* mentions that shrubby species and herbaceous species would hybridize and that the first shrubby species used was *C. bicolor*. This has yellow flowers with a white base. The other shrubby species available to the breeders in the 1830s, apart from the three already mentioned, was *C. thyrsiflora*.

It was not long before cultivars were offered to the public. As early as 1835, H. Major of Knosthorp, near Leeds was offering eleven separate named sorts, including the orange-flowered 'Aurantia' which remained a favourite for a long time. Other colours were, Lemon, Sulphur, Crimson, Dark Claret, Rich Scarlet and Brown.

By 1837 he had a larger range of colours including some bicolors. 'Captain Ross' was red-brown edged with yellow, while 'Lunata' was rich crimson with a yellow edge. In the same year Joseph Plant was offering twenty selected herbaceous types and eight shrubby types. The descriptions sound as though the herbaceous types were very similar to those of today. He did not name his seedlings, but sent them out by number. Thus No. 4 was deep cream edged with rosy purple; No. 10 was copper buff with crimson stripes and spots, while No. 18 was red and yellow with a deep orange red and the whole was 'curiously mottled'. The shrubby varieties sound a good deal more attractive than anything we see nowadays. No. 3 had scarlet-crimson flowers with a rich yellow edge; the colours were beautifully defined, while No. 7 was purple-crimson edged with cherry.

The next year Joseph Harrison was advertising no less than seventy-three named sorts, but it is not clear which were herbaceous and which shrubby. The same year Major offered twenty-four shrubby Calceolarias, several of which sound attractive. For example 'Cassandra' was 'light yellow with an equal portion of the surface of the flower pencilled or grained with chocolate, like oak'. 'Vertumnus' was similarly marked. 'Euterpe' was pure white with a pink blotch and 'Hesperides' was 'French lilac', whatever colour that may be. In 1839 W. Barratt, who had a reputation as a fine grower of other plants came forward with a list of sixty-nine names, but there seems little that is different in the case of the cultivars described.

In recent times the herbaceous hybrids have been crossed with *C. integrifolia,* to form a race of small-flowered coloured Calceolarias; this strain was given the name *C.* X *jeffreyi.* It was again crossed back on *C. integrifolia* to give what would be more compact plants. (I suspect the reference to *C. integrifolia* really refers to the bedding Calceolaria, which, as we have seen, may well be *C. integrifolia* X *amplexicaulis.*) This strain originated at Kew Gardens and was given the name *C.* X *kewense.* All these breeders were doing was to restore the position of the late 1830s.

All the species that were employed in the various hybrids are diploids with 18 chromosomes. Two tetraploid species have been recorded; the hardy herbaceous *C. polyrhiza* and the sub-shrubby white-flowered *C. alba.* This latter was crossed with a garden variety at Messrs Veitch, to give *C.* X *veitchii,* which reached from 3 to 5 feet high and had large corymbs of pale yellow flowers. This was presumably sterile and must be supposed to have disappeared. Another hybrid, raised about 1882 for winter flowering in the greenhouse, was between *C. deflexa* and *C. pavonii,* and was known variously as 'Burbidgei', 'Clibranii' or 'Profusa'. It sounds as though it were an effective plant and it is to be hoped it has survived.

### ❀ *Canna*

The Cannas are denizens of the tropics and subtropics, but are generally found at some elevation so that they are suitable for cultivation in temperate climes. They are used more extensively in countries with a warmer climate than Great Britain, but are grown here as well as elsewhere. They are found wild in South, Central, and Southern North America as well as in China and India. They have been in cultivation for a considerable time; the Indian Shot, *C. indica,* was known as early as 1570.

Hybridization was started in the late 1840s by M. Année, who had been French consul-general at Valparaiso and returned in 1846 with a number of plants. His first hybrid appeared either in 1847 or 1848 and was called 'Annei'.

What were the parents? E. Chaté, who wrote a book called *Le Canna,* published in Paris about 1869, and who knew Année, says that it was *indica* X *nepalensis. Indica* is the common Indian Shot from the West Indies with tall stems and red flowers; *nepalensis* (now known as *chinensis*) is an Asiatic species with deep orange flowers. However Chaté seems to suggest that Année's *nepalensis* was a form of *glauca. C. glauca* is a tall American species with pale yellow flowers. In the *Journal* of the R.H.S. for 1894, J. G. Baker gives a different parentage. He suggests that 'Annei' is *nepalensis* X *glauca.* Since both Chaté and Baker seem agreed that *glauca* is one parent, it is only a question whether *indica* or *chinensis* was the other. The plant was grown not for the sake of the flowers, but for its handsome tropical-looking foliage and most subsequent hybrids were bred for the same reason up to about 1856, when the first flowering hybrids started to supplant them. Some of Année's other foliage hybrids are very difficult to identify. 'Imperator' was *gigantea* X *musaefolia.* Chaté gives *excelsa* as a synonym of the latter species, but I can find no record of this: evidently it was a species with very large leaves. There was also a cross between *musaefolia* and *peruviana,* but I can find no record of *peruviana.* 'Nigricans' was *purpurea* X 'Annei'. *Purpurea* is probably *C. edulis,* which has purple stems and leaves and which is present in almost all the modern dark-leaved cvs. *C. discolor* also has purplish leaves, but the early hybridists complained that it was sterile, and no good as either a pollen or seed parent.

In 1858 Année started using *C. warscewiczii* (introduced only in 1849), again a tall species, with brilliant scarlet flowers and purplish leaves. By crossing this species with 'Annei' he produced a plant with the appalling name of 'Warscewiczioides Annei'. He now turned his hand to breeding for flowers as much as for foliage; he also distributed his stock to nurserymen such as Chaté, Crozy and Sisley at Lyon, Vilmorin and, naturally, Lemoine. The breeding of hybrids was increased. The parentage of 'Annei superba', one of the first of the flowering hybrids, is not known, but his most important crosses used *C. iridiflora,* the only species that is still in cultivation. This is again very tall – the stems can reach a height of 10 feet – but it has long tubular carmine flowers that are very handsome. Année's main hybrids with this were 'Iridiflora hybrida' which was *iridiflora* X 'Imperator' and, most important of all his hybrids, 'Iridiflora rubra' which was *iridiflora* X *warscewiczii.* This was purchased by Kolb at the Munich Botanic Gardens and distributed under the name of 'Ehemannii'. As a result Année's contribution has been rather neglected. Most of the hybrids raised subsequently had 'Ehemannii' as one parent. It was backcrossed to *iridiflora* and crossed with other species such as *glauca.* It is not clear at what time *C. lutea,* a dwarf species, was introduced so as to produce a smaller plant. *C. lutea* normally grows only 2 feet high and this, together perhaps with the equally dwarf *C. limbata,* is the reason for the dwarfness of modern hybrids.

For some reason the early hybrids are known as the Crozy strain, although, as we have seen, Crozy was only one among many working in this field.

A new break came in the 1880s when Herr Sprenger of Dammann & Co. of Naples started using *C. flaccida,* from the south-east United States. This is of moderate height with pale yellow flowers and it brings greater hardiness into the strain, together with what are termed 'orchid-like' flowers.

Sprenger's first cross was *flaccida* X *iridifolia*; subsequently he used *glauca, discolor* (apparently the supposed sterility had been overcome) and '*zebrina*' whatever species that name may cover. It is probably *C. limbata* (syn. *aureovittata*) which shows slight variegation in the leaves and has red and yellow dappled flowers. In the tropics they may have used the white-flowered *C. liliiflora* to produce white flowered cvs., but these are not known in temperate climes although tropical catalogues mention white Cannas. In temperate climes Cannas are usually raised from seed, but in the tropics named cultivars are propagated vegetatively. A tetraploid form of *C. limbata* has been found, but otherwise the species tend to be diploid with 18 somatic chromosomes. They all appear to cross readily and the hybrids are generally fertile, so that the modern Canna is an advanced generation hybrid and various forms have become fixed. In temperate climes no further work seems to be being done with Cannas, but they are being further developed in tropical countries.

## *Clematis*

Clematis hybrids have not perhaps quite so much right to be mentioned here, as some of the species are well-known in cultivation. However no one has seen recently the large-flowered species that gave the hybrid Clematis its attraction. Jackman's book* on the modern Clematis; mentions four species *CC. lanuginosa, patens, fortunei* and *standishii*. The first two are good species, but *C. fortunei* is now considered to be a form of *C. florida,* which has white flowers that tend to be double. *C. standishii* was a variety of *C. lanuginosa.*

The three large-flowered species that form the basis of the modern hybrids are all plants of the Orient. Both *florida* and *patens* had been cultivated for some time by the Japanese before their introduction into Europe and it is doubtful if the wild forms have ever been in western cultivation. *C. florida* is actu-

* G. Jackman and T. Moore, *The Clematis as a garden flower* (London, 1873).

ally a native of China, *C. patens* may well be Japanese and *C. lanuginosa* is Chinese. The first to be introduced was *C. florida* which arrived in 1776. It was possibly introduced by Linnaeus's pupil Thunberg, who wrote the first *Flora* of Japan, published in 1784, and was almost certainly a cultivated form. A double form was soon noted. *C. florida* is the only one of the three large-flowered species with biternate leaves. The plant flowers in late June and early July. *C. florida* can also be distinguished from the other two species by the presence of leaf-like bracts on the flower stalks.

In 1836 the next species followed. This was *C. patens*, introduced by von Siebold, again from cultivated Japanese plants. The wild form is said to be white, but, as known in cultivation, the colour runs from pale lavender to quite a rich violet-blue. The leaves are usually ternate, but occasionally forms with four or five leaflets are observed. *C. patens* flowers in May and June and is the chief parent of all the early-flowering hybrids.

*C. lanuginosa* was introduced by Robert Fortune in 1850. In the wild it is reported to be only 6 feet high, but the hybrids are far more vigorous. The leaves are either simple or ternate and are very woolly on the underside. The flowers vary in colour from white to pale lilac in those forms that have been introduced to cultivation. This is the last of the large species to flower, coming into blossom from July to October.

A form of *C. florida*, known either as *sieboldii* or *bicolor*, with white tepals and purple petaloid staminodes, has persisted in cultivation, but the other forms and the other two species have vanished from our gardens.

The first Clematis hybrid was raised in 1830 and employed none of the species we have been discussing. It was *C.* X 'hendersonii' ('eriostemon') and was created by crossing *C. viticella* and *C. integrifolia*, the latter probably the seed parent. *C. viticella* is a vigorous climber and a native of southern Europe, with small nodding violet flowers. *C. integrifolia* is also a native of southern Europe, but is a herbaceous species with quite large (about one inch long) blue, nodding flowers. The hybrid is a rather weak climber with flowers slightly larger and darker

Plate V

*Fuchsia cordifolia* and seedlings

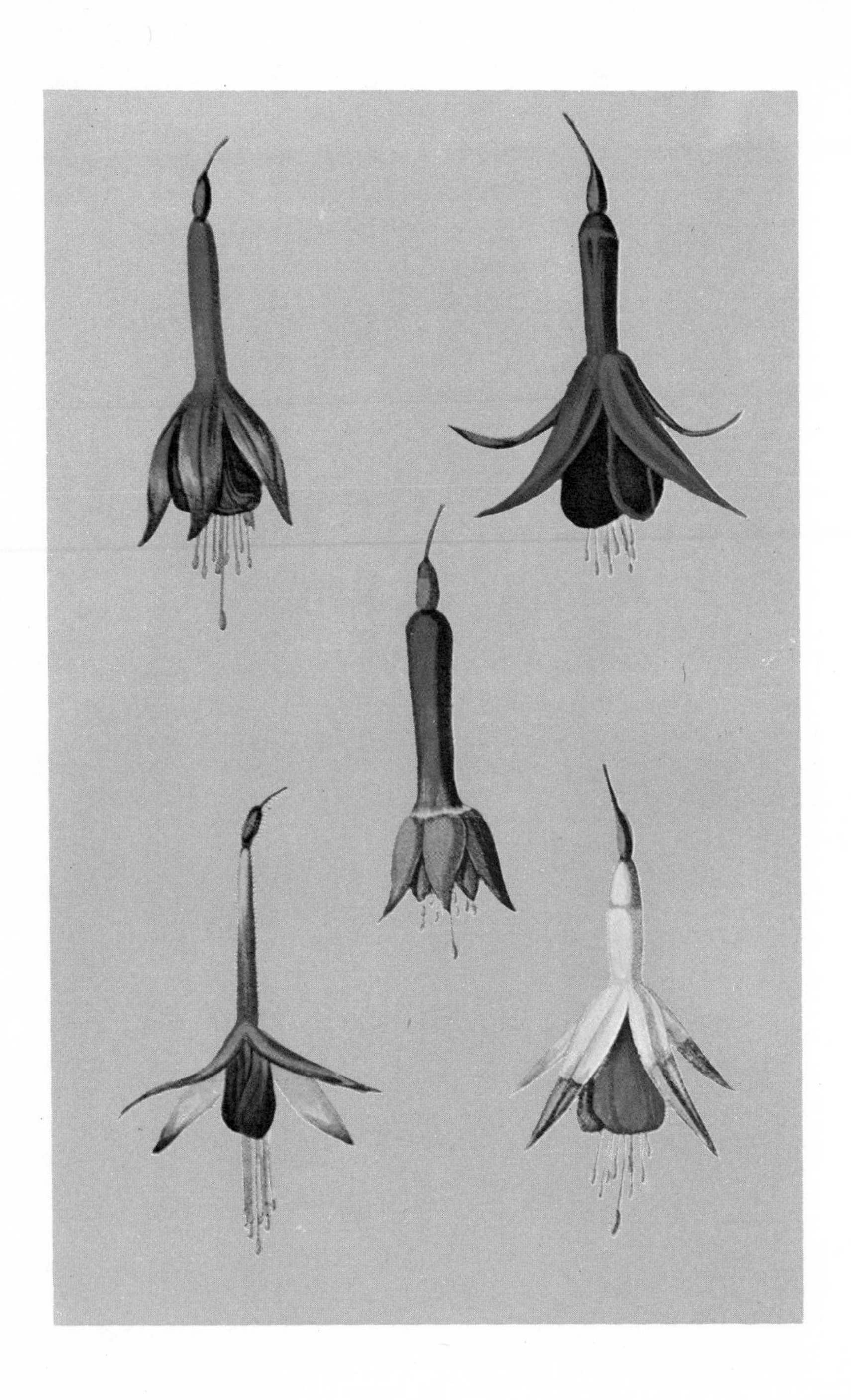

Plate VI
Seedling Calceolarias

than *integrifolia*.

When the large-flowered species were introduced it was found that they were rather shy flowerers and it was only natural that nurserymen should try crossing them with species that flowered better and also with each other in order to induce hybrid vigour (heterosis). The pioneer in this work was Isaac Anderson-Henry of Edinburgh. His first hybrid, dated 1855, was *C.* X 'reginae' and was the result of crossing a form of *patens* called *azurea grandiflora* with *lanuginosa*. His name is better known through the so-called 'Lawsoniana' hybrids (named after the firm of P. Lawson & Son, who distributed them) which were crosses between *lanuginosa* and *florida* var. *fortunei*. One particularly large-flowered white has been named after him as 'Henryi'.

In 1860 Jackman, the well-known nurseryman from Woking, crossed *lanuginosa* with 'hendersonii'. The resultant plants are perhaps the most popular of all the hybrids. They were known as 'Jackmanii' and have large violet velvety flowers. In subsequent crosses they used *C. viticella* in place of 'hendersonii'.

In the early years some nurserymen, including Lemoine at Nancy and Simon-Louis at Metz, were working exclusively with the various forms of *C. patens*. Between 1864 and 1865 a number of hybrids of *lanuginosa* and backcrosses of these hybrids to *lanuginosa* were raised by a Mr Townsend and were greatly admired. Unfortunately they all appear to have been lost. M. Nolle had successes by crossing *C. florida fortunei* with *C. lanuginosa standishii*.

By 1871 a large number of hybrids had been raised, all of which were more floriferous and had larger flowers than the species. Nurserymen proceeded to cross these hybrids and to self them and apparently no further records were kept. In spite of this interbreeding the hybrids tend to come in three distinct races: the early-flowering *patens* group, the mid-season *florida* group and the late *lanuginosa* group.

In 1868 a new species was introduced to cultivation, the North American *C. texensis*. This is a member of the *Viorna* section, characterized by urn-shaped flowers, very different from the open flowers we have been discussing. The plant is tender in England. In the best forms the flowers are a bright scarlet, so it was obviously desirable to try to breed this strain into the large-flowered hybrids. This was attempted, but with only partial success. Such cvs. as 'Ville de Lyon' certainly bring a reddish tinge into the large-flowered hybrids and there is also a shade that is a pale pink as in 'Comtesse de Bouchard' but nothing approaching the scarlet of *C. texensis*. More success has been obtained by crossings with smaller-flowered species such as *C. viticella*. Cultivars such as 'Etoile Rose', 'Gravetye Beauty' and 'Countess of Onslow' have good deep pink rather starry flowers. Most of these do not open widely and, in this country, are not produced very freely, unless grown against a south wall. *C. texensis* is somewhat variable in its colour and the best scarlet forms are not common. The section *Viorna* tends to flower sporadically over a long period, unlike the other sections which give a large display over a short period.

With two exceptions (*CC. mandschuria* and *paniculata*) all the Clematis species are diploids with 16 chromosomes, so there is no obvious reason why other species should not be bred into the large-flowered strain. However since 1880 very little controlled breeding appears to have been done and the introduction of a new Clematis is liable to be due to the lucky chance of a seedling from some existing hybrid. There may well be some hybridity bar, but it would seem to be worth while trying to get the yellow colour of such species as *C. tangutica* into the large-flowered hybrids; and the great vigour and floriferousness of species such as *spooneri* and *montana* could also be valuable introductions into the strain. The large-flowered Clematis seem to have been developed over a comparatively short period and then left, as though perfection had been achieved once and for all. This, as all gardeners are well aware, never happens.

## ❀ *Dahlia*

The history of the Dahlia is a fascinating story of detection on the part of the cytologists. The species that are considered to be the parents of the garden Dahlia have 32 chromosomes, so one would imagine that they were diploids with a haploid count of 16. However, when the behaviour of the chromosomes at meiosis was studied, it was found that the species were all tetraploids and that originally there must have been an ancestral species of Dahlia with a haploid count of 8 and a somatic count of 16. The cultivated Dahlia has 64 somatic chromosomes and is therefore octuploid. It is this high degree of polyploidy which gives the Dahlia its great variability and the cultivated form has been given the name of *Dahlia variabilis*.

The species come from Mexico and it would seem that the original hybridization was done by the natives. Double flowers were, apparently, reported by travellers in Mexico in 1575 (Lawrence and Crane, p. 87) and one is inclined to wonder if they could be a relic of Aztec gardens. There is little evidence that any deliberate breeding was done by European breeders until comparatively recently.

The first plants were introduced in Madrid in 1789 and given the names of *D. pinnata* and *D. coccinea*. The first species had purple semi-double flowers and the second had single red ones. The typical *D. variabilis* appears to have been depicted in the seventeenth century in the work of Francesco Hernandez and it was also described in 1787 by Thierry de Menonville. Seed of the two Madrid species was sent to England by the Marchioness of Bute, wife of the British Ambassador at Madrid, but they were lost and do not seem even to have flowered, although kept for two or three years. However in 1802 *D. coccinea* was flowered by a nurseryman John Fraser, who had obtained seed from France, and in 1803 it was figured in the *Botanical Magazine*. The plants were grown more easily in Spain and France, but seem to have shown little variation.

In 1804 Humboldt sent seeds from Mexico to Paris and Berlin and these seem to have been from *D. variabilis* as we know it today. The seeds showed great variability and by 1806 the Berlin Botanic Gardens were able to flower fifty-five cultivars, which were either single or semi-double. In 1808

Hartweg bred the first perfect double. This is said to have been a form of *D. coccinea,* with scarlet flowers of the pompom type. The next year the first white flower – a single – was recorded. In 1814 M. Donckelaar raised some larger-flowered double varieties at the Botanic Garden in Louvain. Fourteen years later he raised the first race of dwarf plants, the ancestors of the modern Coltness strain. Donckelaar's plants were imported into England and it is from 1814 that serious Dahlia cultivation began in this country.

By 1818 practically every colour known to us today had been obtained. This is a very rapid increase, when you consider that Humboldt's seeds had been received only fourteen years previously. Only four years after Donckelaar's plants had been imported, British catalogues listed eighteen doubles and over a hundred semi-doubles and singles. It was around this time that Count Lelieur produced the first so-called Fancy Dahlias. These were bicolors, either with striped petals or with petals tipped with a different colour.

'Waverley', the first pure white double appeared in 1821. Such flowers do not seem to occur very frequently. Even nowadays there is only a limited choice of double whites. In 1830 the 'Anemone-flowered' type appeared. In this the disk florets are elongated and enlarged, so that the flower is thought to resemble a double Anemone, although it is a long time since the single anemone with petaloid staminodes has been in cultivation. The modern double Anemone has several rows of tepals and generally normal stamens. Varieties were now increasing to a ludicrous degree. In 1831 a Swiss amateur had 1,500 varieties in his garden and ten years later Joseph Harrison had 1,200 different doubles at his nursery.

Although the pompom Dahlia had appeared in 1808, it was a small flower on a large plant. Between 1845 and 1850 this was developed so as to match the small flowers with a correspondingly small plant.

In 1872 a certain Mr J. T. Van der Burgh, who lived near Utrecht, received a box of plants from Mexico. Most of the plants had perished on the journey, but among the few survivors was a small Dahlia tuber. This eventually produced a double scarlet flower with long, narrow petals with recurved margins. It was given the name of *Dahlia juarezii,* but it is now thought not to be a true species but a hybrid derivative from *D. popenovii.* It seems curious in this case that so striking a form had not been introduced before. A good stock had been worked up by 1874 and it was distributed by Anton Roozen & Sons. This was soon bred into the existing Dahlias and gave rise to the so-called Cactus Dahlia. It obtained this name as the flowers were supposed to resemble the flowers of the *Cereus* cactus. When the Cactus was crossed with the large globular Show Dahlia the results were called Decorative or Paeony Dahlias. They were very little regarded at first, though they subsequently became the most popular type of all Dahlias; but it was only at the beginning of this century, that they were developed seriously.

In 1880 Alfred Salter brought to the R.H.S. what he called *D. coccinea,* a single scarlet; and at the same time a single yellow form appeared. This started a short-lived enthusiasm for single Dahlias, which, one presumes, had only 32 chromosomes. In 1890, or slightly before, Mr E. J. Lowe crossed a single Dahlia 'Stella bianca' with *D. merckii.* This latter species is still occasionally to be found in cultivation. It has rather short lilac ray florets, which are pointed at the tips. It is unique among the genus for having two extra pairs of chromosomes, giving the atypical count of 36. Mr Lowe's cross, which was introduced by Messrs Dobbie in 1891, was first known as Single Cactus, but subsequently as the Star Dahlia.

What appears to be a typical plant of Victorian gardens, the Collarette Dahlia, which has an inner ring of ray florets of a different colour from the outer ring, was not developed until 1900. It reappeared on the show bench in 1966 after a long absence. As a general rule the Dahlia is self-sterile. In *The Dahlia: Its History and Cultivation* by a number of authors, published by Macmillan in 1903 (the first edition was 1897, but I have been using the 1903 revised edition) there is an account of an experiment to prove that the Dahlia was self-fertile, but it does not appear to have been very well planned and there seems no reason to doubt the generally accepted opinion. As a result, although the raising of new varieties is easy, all cultivars can only be propagated by cuttings or division of the tubers. Whether it would be possible to breed in further species is uncertain (although the resultant plants should be hexaploid and therefore fertile). It is also doubtful whether it would be desirable; apart from the enormous *D. imperialis*, with stems up to 9 feet in height and white, red-centred flowers, there are no other species in cultivation and the genus seems little known as a whole. *D. excelsa*, which needs greenhouse shelter in this country, grows 20 feet high and it might be possible to breed a race of tree Dahlias for subtropical countries. The scheme has its attractions, but whether it would be commercially rewarding is somewhat doubtful and this uncertainty has presumably prevented the cross being attempted.

As things are it seems unlikely that any great variation may be expected in the Dahlia in the future. Possibly the range of colours may be extended, but that seems to be the limit of what we may expect, unless some new species are collected and bred into the strain. This, however, presupposes that there are some desirable new species to be collected and there is no evidence of this.

### ❀ *Delphinium*

The race of garden Delphiniums seems rather to have arisen by accident, and there is very little information available as to its origins. It would seem that the main species are *DD. cheilanthum, elatum, exaltatum, formosum,* probably *grandiflorum* and its var. *chinense*. Other species that have been used include *brunonianum, tatseniense, cardinale, nudicaule* and *zalil*. The four first-mentioned species are the principals. *D. cheilanthum* is a species from Siberia and China, growing from 2 to 3 feet high according to the R.H.S. *Dictionary*, rather taller according to Bailey's *Hortus*. According to Bailey the so-called Belladonna Delphinium is *D. cheilanthum* var. *formosum*; others have suggested that it is a hybrid of *elatum* with *grandiflorum* var. *chinense,* while most writers are unable to suggest a parentage for this at all. 'Belladonna' is a hexaploid and more or less sterile, which suggests that Bailey's attribution is doubtful.

*Delphinium elatum* has a wide range from France to Siberia, although it was from its furthest station that it was introduced to cultivation in 1597. This reaches a height of 6 feet and is very similar to the garden Delphinium in outward appearance. Of course the flowers are smaller and not semi-double, but its appearance is more like the cultivated plant than the other putative parents. The cultivated plant is a tetraploid with $2n=32$ while most wild species are diploids with 16 chromosomes in the somatic cells. There are a few natural species that are tetraploids including *D. carolinianum* and *D. speciosum* (*caucasicum*), but there seems no reason to suppose that they have been employed in breeding.

*Delphinium formosum* is a Caucasian plant with branched stems reaching to about 3 feet and seems to produce white forms more easily than the other species, so that it is probable that the white cvs. have a large share of *D. formosum* var. *album* in their make up.

It is probable that the ordinary Delphinium is the result of the hybridization of these three species. When tetraploid forms arose these were selected for their larger flowers and more robust habit and the original diploid plants were superseded.

*Delphinium grandiflorum* is a quick growing species from Siberia and the western parts of the U.S.A. with its var. *chinense* in China. This is a shortish plant, usually only about 18 inches to 2 feet high, with the flowers more widely separated on the branched spikes than in the three previous species. It is easily recognizable by its leaves divided into linear segments and it is very rapid in growth. It may, indeed, be treated as a hardy annual. It shows a good range of colours in varying shades of blue, and white forms are also known. Similarly single and double flowers are present in nearly equal measure in the cultivated strains. From appearances one would assume that the Belladonna delphinium had *D. grandiflorum* in its ancestry.

It is a triploid and was sterile until 1902 when a hexaploid form appeared.

*D. brunonianum* is a dwarf species from central Asia with rather dark violet flowers. Assuming that G. A. Phillips is correct when he states in his *Delphiniums, their History and Cultivation* (London, 1932) that this species was used, it is presumably responsible for the purple and violet colours. However, unless a tetraploid form has been discovered, one would expect these cultivars to be sterile.

*D. tatseniense* is a dwarf Chinese species which was used by Lemoine to breed a race of Delphiniums that would be suitable for bedding. They bred two cultivars 'Colibri' (1914) and 'Libellule' (1913). For these, the wild diploid form of *D. elatum* was used and it seems a pity that these attractive dwarf plants did not prove more popular.

In the 1930s a German grower, L. Lindner of Eisenach, crossed the diploid form of *D. elatum* with *D. cashmerianum*. The latter species has azure flowers in corymbs rather than in a spike, so that the inflorescence of these hybrids was less erect and therefore less liable to wind damage than the more conventional hybrids. Possibly because of this, the plants seem not to have achieved much commercial success.

There are two scarlet-flowered species of Delphinium, *D. cardinale* and *D. nudicaule*. The first is a tall species from the Californian desert and has very good resistance to mildew. *D. nudicaule* is also Californian, a smaller plant with colour forms ranging from bright yellow, through scarlet to purple and is not particularly resistant to mildew. In spite of this it was the latter species that Ruys attempted to cross with the garden Delphinium in the 1920s. As might be expected, the cross appeared to be quite impossible. They attempted to use *D. nudicaule* as the seed parent and as the pollen parent with no apparent result. After the experiment had been abandoned it was found that one seedling from a seedbed of *D. nudicaule* appeared to be a hybrid with *elatum*. This was indeed the case and the resultant second generation cv. 'Pink Sensation' made a furore. It is a Belladonna type with pink flowers. The plant proved to be a tetraploid with

41. *Canna warscewiczii*

42. *Canna iridiflora*

43. *Clematis patens*

44. *Clematis texensis*

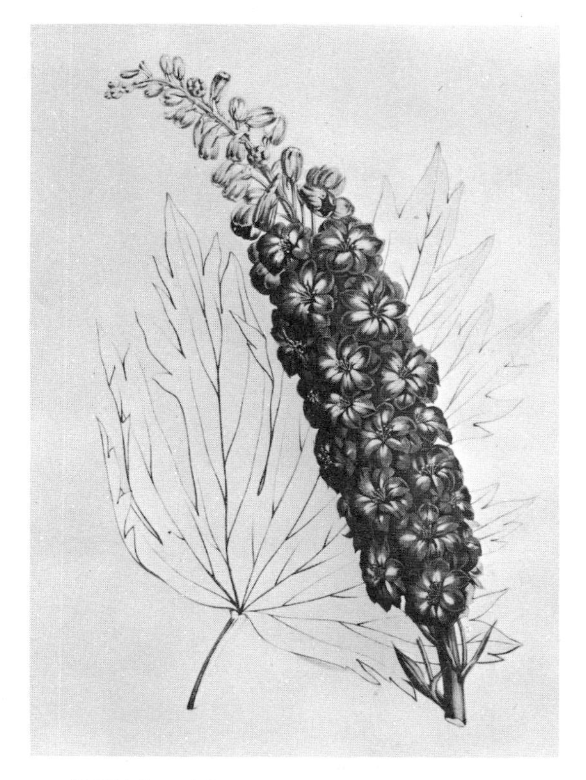

45. *Delphinium elatum*

46. *Delphinium formosum*

47. *Gladiolus Colvillei* hybrids

48. *Gladiolus psittacinus*, 1830

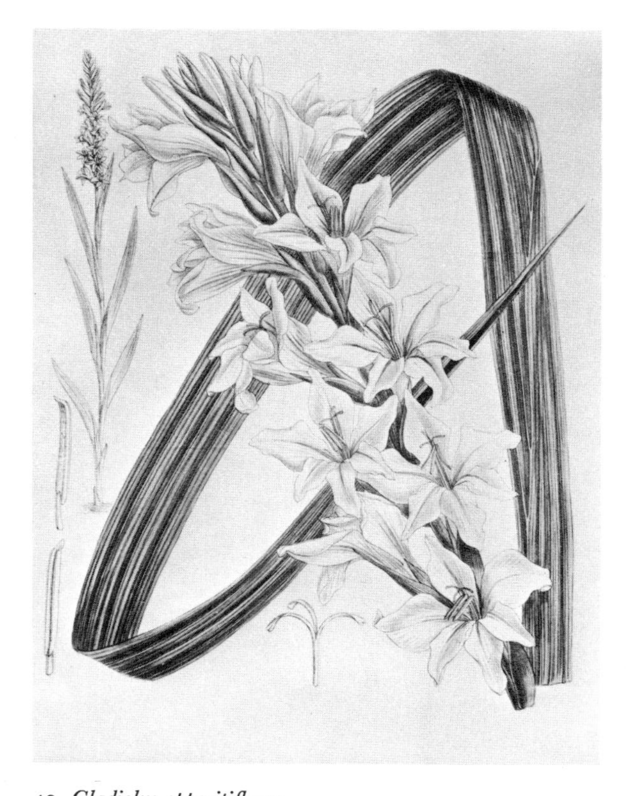

49. *Gladiolus oppositiflorus*

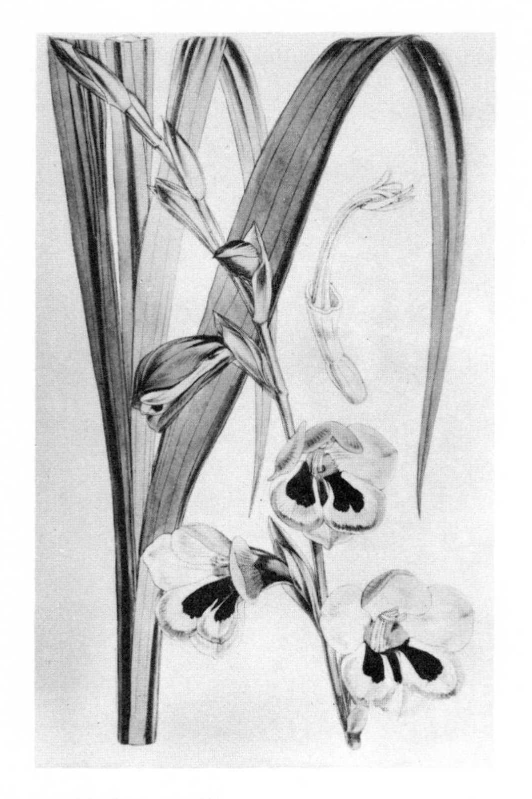

50. *Gladiolus papilio*

51. *Gladiolus saundersii*

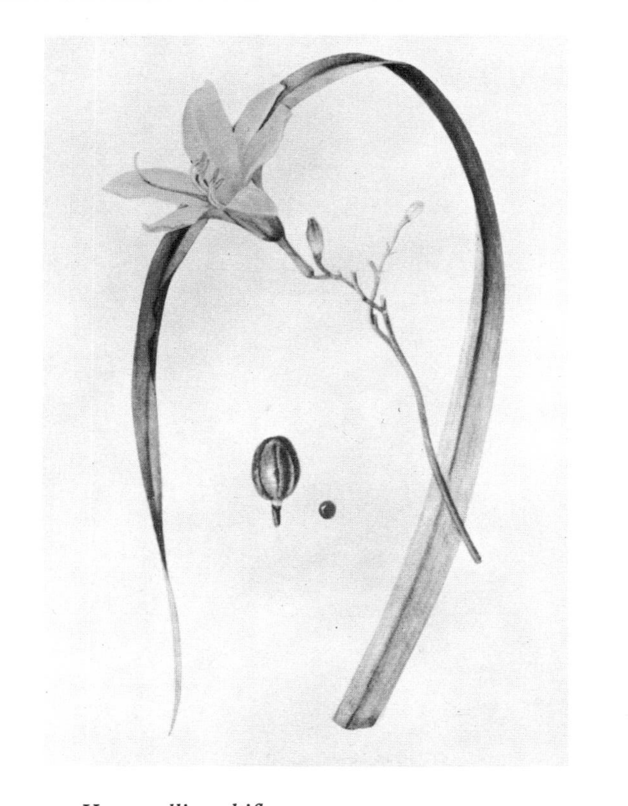

52. *Hemerocallis multiflora*

53. *Hemerocallis forrestii*

54. *Hippeastrum reginae*, 1799

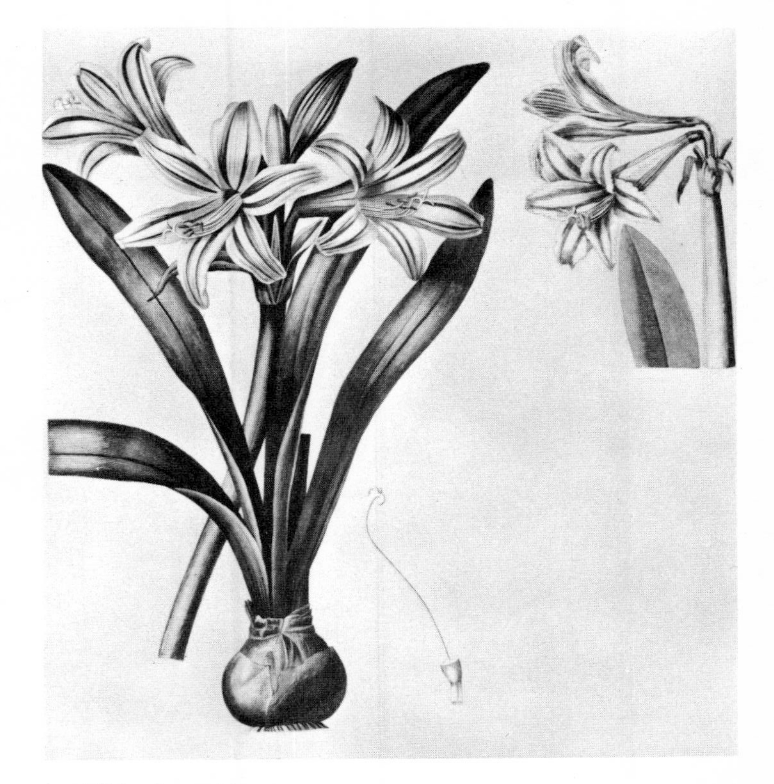

55. *Hippeastrum vittatum*

56. *Hippeastrum reticulatum*

57. *Hippeastrum solandrifolium*, 1825

58. *Pentstemon cobaea*, 1836

59. *Pentstemon hartwegii*

60. *Petunia axillaris*

61. *Petunia violacea*

62. *Phlox paniculata*, 1817

63. *Phlox carolina*, 1810

64. *Verbena platensis*, 1838

65. *Verbena phlogiflora*, 1836

66. *Verbena peruviana*, 1834

67. *Verbena incisa*, 1838

68. *Deciduous Azalea*

32 chromosomes, and this accounts for its fertility.

In 1953, Dr R. A. Legro started breeding experiments at the Horticultural laboratory at Wageningen in Holland. After natural pollination between the wild species and various white cultivars had been proved impossible, it was resolved to obtain tetraploid forms of *nudicaule, cardinale* and *zalil.* This last is a yellow-flowered species from Persia. It was comparatively easy to obtain tetraploid forms of *cardinale* and *nudicaule* with colchicine, but it took seven years and the treatment of five thousand seedlings before tetraploid forms of *D. zalil* could be bred.

The creation of these tetraploid forms did not, however, solve the problem. Used as a pollen parent, the results with *cardinale* and *nudicaule* were nil, while *zalil* produced attractive plants that were unfortunately completely sterile. Although tetraploid scarlet pollen on garden forms had no result, the cross of the garden cvs. on the scarlet plants did produce seeds and the offspring were fertile, but the results even in the second, third and fourth generations were disappointing.

Dr Legro now had the brilliant idea of crossing *Dd. nudicaule* and *cardinale* on the one hand, *Dd. zalil* and *cardinale* on the other and the resultant seedlings were treated with colchicine to give allotetraploids, which might be expected to be highly fertile. Dr Legro refers to the first cross as *D.n.c.* and the second as *D.z.c.* Used as a pollen parent, *D.n.c.* was ineffective, but the reciprocal cross, with *D.n.c.* as the seed parent, proved highly successful. The pollen parent was a white cultivar 'Black and White'. This is semi-double but the resultant seedlings were nearly all single, in colours ranging from blue to violet. However in the second generation a number of different colours appeared including one plant with semi-double red flowers that were two inches across. The most outstanding of these second generation crosses was 'Orange Beauty', which resembled the well-known garden delphinium, but had orange flowers. Unfortunately it proved self-sterile, but it has good pollen and will set seed from pollen from other cvs.

*D.z.c.* proved useless as a seed parent, but would set seeds when used as a pollen parent. However, once again the resultant plants produced no pollen and would not accept any. However, for no clearly explained reason, it was found that if *D.z.c.* was crossed with the second generation *D.n.c.* X 'Black and White', not only could the crosses be made in either direction, but the resultant plants were fertile. The same results were obtained when the *D.n.c.* hybrids were crossed with the tetraploid *D. zalil.* The object of all this work was to produce plants like the large cvs. but with red, orange and yellow flowers. This has to a large extent been accomplished. It is true that so far we have not the massive spikes of the giant cvs. but that, to my eye, is an advantage. It is safe to say that Legro has transformed the Delphinium as we knew it, although the plants seem to be taking their time to come into commerce. They were first exhibited in 1962.

Dr Legro's latest exploit has been to incorporate two perfumed Delphinium into the strain which has been named the University Hybrids. These species are the white, strongly-scented *D. leroyi,* from the mountains of Central Africa, and the pale blue, less strongly-scented *D. wellbyi* from Abyssinia. Primary hybrids showed that this fragrance was, somewhat unusually, a dominant characteristic genetically. Dr Legro therefore crossed *leroyi* with *wellbyi,* treated the resultant seedlings with colchicine and then crossed this tetraploid cross with *elatum* cultivars. It was found that the *elatum* cvs. had to be used as pollen parents: the opposite cross produced no pollen. The resultant plants have all been fragrant. Great interest has also been roused by the cross *zalil* X *leroyi,* which has given rise to a large-flowered fragrant hybrid, in colours ranging from creamy-yellow to pale pink. If this can be bred into the large-flowered *elatum* cvs. we may obtain a large yellow fragrant Delphinium. The cross *leroyi* X *cardinale,* when made tetraploid, has been found to combine with *elatum* cvs. in both directions. It has been possible to use *elatum* as a seed parent and the resultant hybrids have shown marked hybrid vigour. Since the original cross combines fragrance with mildew-resistance and both characters appear dominant, we may expect great results. (For further details consult G. T. Cairncross, *Gardeners' Chronicle*, June 28, 1967.)

### ❀ *Freesia*

These delightful, fragrant cormous plants of so many different colours have all been bred from two species. The first, introduced to cultivation in 1815 as *Tritonia refracta* and now known as *Freesia refracta* is a small plant with white or yellow flowers. The other species is similar but has pink flowers; this is *F. armstrongii* and was not in cultivation until 1898. It is from the crossing of these two species that have come all the reds, mauves and violets, as well as the whites and yellows. At some period in cultivation the strain became tetraploid and this may in some way explain the great range of variation in the colours. A few years ago double cultivars appeared, but they do not seem to have been much heard of recently. They resembled gardenias rather than freesias and seemed to be an important new break and it will be interesting to see how they are developed in the future. They appeared to be so very double, that it looked probable that both stamens and stigmas had become petaloid, so that propagation could only be by offsets. Freesias usually produce large numbers of these, but possibly this was not the case with the doubles.

### ❀ *Gladiolus*

All the garden hybrids have been raised from South African species and from very few of them, although there are a vast quantity to choose from. This would seem to be because size of flower has been preferred to delicacy and fragrance. Anyone who has seen at various R.H.S. shows the delightful, fragrant hybrids raised by Professor T. Barnard, must wish that they were available commercially; and it is interesting to see that work on similar lines was being done, principally by Dean Herbert, in the 1820s (and also by Amos Perry in the 1930s). Sweet's *Hortus Britannicus* records the following hybrids:

*blandus* X *cardinalis, tristis* X *hirsutus, tristis* X *blandus, recurvus* X *tristis, versicolor* X *blandus, recurvus* X *blandus* and two hybrids between (*blandus* X *cardinalis*) with *recurvus* and *hirsutus.* Of all these species only *G. tristis,* with greenish-yellow night-fragrant flowers is currently in cultivation. The other species cannot be purchased, although they contain some delightful flowers. The only one of the early hybrids to survive is 'Colvillei' raised in 1823 reputedly from *cardinalis* X *tristis* although the influence of the pollen parent, *tristis,* is barely apparent and the plant looks very like a smaller *G. cardinalis.* From 'Colvillei' the 'nanus' strain was bred. Both *Gg. cardinalis* and *tristis* are diploids ($2n = 30$), but 'Colvillei' exists in diploid and triploid forms, while the 'nanus' race adds a tetraploid. The plants are useful for their early flowering, but have not been much developed.

The large-flowered race comes from a series of hybrids. In 1842 the first of the 'Gandavensis' hybrids were raised in Ghent. The parents were *G. psittacinus* and *G. oppositiflorus. G. psittacinus* is a tall species with a hooded flower that is red with a yellow centre; *G. oppositiflorus* is up to 4 feet high, with white flowers marked with a violet stripe. It is unique among the Gladiolus for its opposite-flowered spike. Probably all the modern white Gladiolus derive from this species. *G. psittacinus* was crossed with *G. cardinalis* to give the 'Brenchleyensis' hybrids. These were almost all red and yellow, but some had the typical *cardinalis* blotches.

Introduced from 1880 onwards, Lemoine's strain incorporated two new species. *G. purpureo-auratus,* which can still be met with occasionally, has hooded yellow flowers with purple-blotches. It is a hardy plant and a rapid increaser. *G. papilio* is similar but with a pale mauve ground colour and it is from this that all the lavender and violet cvs. derive.

At about the same time, Max Leichtlin in Baden-Baden was trying to breed in *G. saundersii.* This species has very large scarlet flowers with a white blotch in the throat. It does not have many flowers to the spike and these tend to open singly; these undesirable qualities were transmitted to the first generation hybrids, so that Leichtlin's cvs. were not very popular. In addition *G. saundersii,* which comes from high mountains, is intolerant of drought and the corms do not store well. Leichtlin sold his stock to a Long Island nurseryman, who developed them and then sold them to J. L. Childs, who launched them as the 'Childsii' strain. At the same time Lemoine had used *G. saundersii* in his Lemoine hybrids, calling the resultant cvs. the Nancy or 'Nanceanus' strain. Another handsome mountain species that proved extremely difficult in cultivation was *G. cruentus.* Dr W. Van Fleet employed on the 'Childsii' strain and, although most of the resultant hybrids inherited the difficult constitution of *G. cruentus,* one, 'Princeps', had the large size flowers of *G. cruentus* with the strong constitution of the 'Gandavensis' strain. The majority of giant-flowered Gladioli may have 'Princeps' in their ancestry.

A species not from South Africa, but from mountain heights in tropical Africa, is *G. quartinianus.* This is a tall, strong-growing plant, reaching up to 4 feet in height and of easy culture. The red and yellow hooded flowers are not individually large, but are numerous on the spike. Both Kunderd of Indiana and Van Fleet used this species to combine with the others, so that the modern Gladiolus is comprised of many species.

Last of all to be used, but perhaps the most important, is *G. primulinus** which was not introduced until 1889. This is confined to a small area around Victoria Falls and was nearly exterminated by over-enthusiastic collectors. It has a clear yellow flower with a pronounced hooded shape and when bred into the various large-flowered hybrids has produced smaller but more attractive flowers, and has enabled various delicate shades to be added to the rather crude colours of the original hybrids. *G. primulinus* is a denizen of the tropics, but the *primulinus* hybrids appear as hardy as any other hybrid Gladiolus. Primulinus hybrids combined with the ruffled flowers that were developed by Mr Kunderd have given rise to the so-called 'Butterfly' strain, which has small ruffled flowers. First generation *primulinus* crosses tended to open only one or two flowers

* Now known as *G. nebulicola.*

at a time and these drooped somewhat, so that the display was not very striking. Advanced generation hybrids have done away with these defects.

There appears no reason why other species should not be added to these hybrids and the dwarf plants can certainly be improved by hybridization. The published chromosome counts of the various species are somewhat confusing. A number of pentaploids are recorded, but apparently the plants used were cultivated and the wild species are, presumably, viable. Possibly the chromosomes do not pair regularly at meiosis, but I have been unable to find any accounts of this. Since many of the pentaploids are known to have been used in breeding, we may assume either that the published figures are atypical or that the division at meiosis is unusual.

### *Hemerocallis*

Although the breeding of hybrid Hemerocallis was started by George Yeld about seventy years ago, it is only in the last thirty years that they have been regarded as really valuable garden plants. The present race is the result of multiple hybrids between the thirteen species known, although it is doubtful if much use has been made of either *H. forrestii* or *H. exaltata.*

Hemerocallis are all natives of Asia, chiefly of China and Japan, although the triploid clone of *H. fulva,* which is known as 'Europa', has been known in the west since 1576. It presumably spread westwards from China and has survived to this day. A diploid form under the name *H. disticha* was received from China in 1798, but does not appear to have persisted in cultivation and most modern diploid forms of this species are of recent collecting in China and Japan. Forms from Kuling, with flowers that are pink or red, have been given the varietal name *rosea.* The plant was not generally available until 1930 and it is this variety that has so enriched the colouring of the later hybrids.

The other orange-flowered species, *H. aurantiaca,* is not well known in the wild. It is said to grow around Mount Ibuki.

However, the wild plants differ somewhat from the cultivated plant that Baker named *H. aurantiaca* in 1890. In spite of its name the plant has a tinge of fulvous-red in its orange and thus differs from the orange-yellow species. It is possible that the plant may be of horticultural origin as only one clone seems to exist. The plant is self-sterile, although it can be used as a pollinator, and it will set seed to pollen from other species. Somewhat similar, but larger and with no trace of fulvous in its colouring is the clone known as *H. aurantiaca* 'Major'; this is similarly self-sterile, and it has been suggested that it is an accidental hybrid.

It is from these two species, *HH. fulva* and *aurantiaca,* that the dark cultivars have been bred, but it is rarely that the parentage is a simple cross between two species. Thus, for example, the cultivar known as 'Bijou' is a cross between a hybrid of *HH. aurantiaca, flava* and *fulva* with *H. multiflora.* A similar situation now exists in most of the modern cvs. and both the original species and the first interspecific hybrids are now very rarely seen. It may be as well here to list the other species, one or two of which are still to be found in catalogues. *H. flava* has been known in Europe since 1570 and is a most handsome species, with clear lemon-yellow flowers. It is one of the first of the species to flower and, unlike most of the family, it is self-fertile. It is followed after about ten days by *H. minor,* which is smaller in all its parts, but not dissimilar. However, the outside of the tepals are tinged with red. Equally early is *H. dumortierii* with orange-yellow flowers. The scapes tend to be shorter than the leaves but project at an angle so that the flowers are clearly visible. The backs of the tepals are reddish. Superficially very similar is *H. middendorfii* which has taller scapes that overtop the leaves. The flowers open more widely and the shape of the capsules is very distinct from *H. dumortierii.* Those of the latter species are round, while those of *H. middendorfii* are ellipsoid.

Three dwarf species have been introduced to cultivation comparatively recently. They all follow on from the early-flowering group mentioned above and have all been used to create dwarf hybrids, although they are very rarely seen. The species are *H. forrestii* (1912), *H. nana* (1916) and *H. plicata* (1923). The original description of *H. forrestii* stated that the flowers were fulvous, but this character has not been found in any cultivated plants. *H. plicata* does show a slight fulvous tinge, while *H. nana* has the same orange-yellow as *H. dumortierii.*

The next species to flower, *HH. aurantiaca* and *fulva* have already been discussed and so we come to *H. citrina.* This is unique among the genus for its habit of opening in the late afternoon and is predominantly night-flowering, a character it transmits to first generation hybrids. It is a very vigorous and attractive plant, with very pale lemon-yellow flowers, which are borne in great profusion. Dr A. B. Stout in his *Daylilies* (Macmillan, New York, 1934), a book to which this section is greatly indebted, states that he has seen a scape bearing 64 buds. In spite of its night-flowering, it was used for hybridizing by Willy Müller at Naples.

*H. thunbergii*, a Japanese species, is a late-flowering plant with clear yellow flowers, which usually appear in August, although more recent gatherings have been observed to flower later. It has been widely used in hybridization.

*H. multiflora* was described as recently as 1929. This is the latest flowering species known. The scapes are branched and continue flowering over a long period, generally from August until the frosts come. The colour is pale orange and the individual flowers are not large, neither is the plant particularly showy. However it is a very useful parent, as it transmits its habit of carrying many flowers, together with its long flowering season, to its progeny, while it does not seem to inhibit the production of larger flowers.

The last species to be introduced, *H. exaltata* (1934) is a tall plant with scapes reaching 4 feet. The pale orange flowers are quite large and appear in late June and July. The plant was presumably in cultivation at one time, but I have never seen it, nor have I seen it offered. It has, however, been used in breeding.

The pioneer work in hybridizing Hemerocallis was done by

George Yeld, who obtained a Cetificate of Merit for 'Apricot' in 1892. Unfortunately Yeld lost many of his records and it is not certain if this was *flava* X *fulva* or *flava* X *middendorfii*. Yeld was followed by Amos Perry, who worked principally with *fulva* as the pollen parent. Willy Müller worked extensively with *H. citrina* and produced a race of variously coloured night-blooming plants, which were, presumably, popular in Italy. However it is to Dr Stout that the credit must be given for the transformation that the plant has undergone in the last thirty years. He seems to have been the first to combine more than two species in a hybrid. We have already mentioned his 'Bijou'. Another of his early successes was 'Nada' which is *H. nana* pollinated by 'Europa', a plant bred from *HH. flava, aurantiaca* and *fulva*. 'Europa had dark red flowers and 'Nada' has 'rich Morocco red' flowers 4½ inches across, but the plant was only a foot high.

The development of the Hemerocallis is the most spectacular creation of a new garden race within a comparatively short time. There is no reason, moreover, why we should not expect further developments. The various species come into flower from May until the autumn frosts and it should be possible to breed hybrids that would have an equally long flowering season. At the moment the majority are June and July flowerers and it would seem that there is a lack of choice early and late-flowering cultivars. And there seem to be few dwarf plants about at present; this particular line has not been pursued as much as might be expected. In the latter years of the last century double and variegated-leaved forms of *H. fulva* were imported from Japan. These are triploid and the stigmas are aborted. However Dr Stout says that some viable pollen may be obtained from the double flowers. He also reports that some forms of *H. aurantiaca* seem to have a remontant quality and this could well be valuable in the future. Reading Dr Stout's book, one has the feeling that many of his predictions have yet to be fulfilled. What, for example, became of the unnamed night-blooming species with scapes 7 feet high? These giant cultivars have yet to be bred. Colour seems to have progressed little since 1934, except perhaps in a very pale lemon-yellow; perhaps we cannot hope for many new breaks here, unless some new species are discovered. Obviously a pure white Hemerocallis would be very attractive, but white is unknown in the various species. As it is we have a range from very pale yellow through various yellows, oranges and reds to dark maroon and mahogany, an astonishing range. The species will tolerate a wide range of climate from our own temperate one to one that is semi-tropical, although the deciduous species are unhappy under semi-tropical conditions. Since Dr Stout's day the raising of new cultivars has been mainly in the U.S.A.

### ❀ *Hippeastrum*

These seem to be the most involved hybrids that we have yet considered. Growers appear to have hybridized them almost from the moment of their introduction and hybrids were then crossed with hybrids. These are set out in some detail in Sweet's *Hortus Britannicus* of 1839, although it is not easy to identify all his species. What is considered to be the first hybrid is 'Johnsonii' which is *H. reginae* X *H. vittatum*. This was used both as a pollen and as a seed parent in numerous other crosses. Thus 'Sweetii' is *reticulatum* X 'Johnsonii', while 'Colvillii' is the reciprocal cross. 'Formosa' is *reticulatum* X *reginae* and so is 'Gloriosa', the var. *striatifolia* of *reticulatum* having been used in the first cross. The species that were employed seem to have been, *HH. reticulatum, reginae, equestre, vittatum, rutilum, solandriflorum, stylosum, psittacinum*; for some reason *H. aulicum* is not included, although there is reason to suppose that it was used subsequently, notably in the hybrid called 'Ackermannii'.

Of these species, *H. equestre* is comparatively dwarf with scarlet flowers and *H. rutilum* is not so very much larger, but has several colour variations. *H. rutilum* var. *acuminatum* (*pulverulentum*) has a grey bloom on the reverse of the petals, *crocatum* has saffron-yellow flowers and var. *fulgidum* is a dark var. crimson, large-flowered type. In Sweet's *Hortus* all these varieties appear as separate species.

*H. reginae* is slightly larger and has bright red flowers with a white star in the throat of the flowers. *H. reticulatum* has a prominent white midrib to the leaves (var. *striatifolium*) and has a white or pale pink flower netted with crimson veins. *H. vittatum* has white flowers striped with red. *H. solandriflorum* has very large flowers of a greenish-white colour. *H. stylosum* is a comparatively small plant with brownish-pink spotted flowers. *H. psittacinum* has flowers in green and scarlet stripes, while *H. aulicum* is deep crimson with a purple-rimmed green blotch in the centre of the flower.

The discovery and introduction from Peru in 1870 of the large-flowered scarlet *H. leopoldii* brought about a transformation of the hybrid strain. *H. leopoldii* has flowers some 7 inches across, and this huge size was transmitted to its hybrids, but they did not have the variety of shape and colour that had previously existed. As so often happens in plant breeding, size was preferred to grace and delicacy. By 1880, the 'Amaryllis' as they came to be called, had become a race of huge-flowered, predominantly scarlet- or red-petalled plants, with a certain amount of white, which could sometimes be striking, but at other times was restricted to the throat of the flower. Occasionally striped flowers showed that *H. vittatum* was still being employed, but the scarlet colour of *H. leopoldii* tended to be dominant over the more delicate colours of some of the other species that had previously been employed.

It is thus probable that at least ten species have entered into our modern 'Amaryllis'. Of recent years the hybrids have been crossed back on to *H. equestre* in an endeavour to produce a smaller and more floriferous plant. Although the hybrids are amongst the most imposing of flowers, some of the species sound as though they might well be more attractive and *H. solandriflorum,* in particular, sounds outstanding. Little seems to be known of the cytology of Hippeastrum. The haploid number is 11 and the only two that appear to have been counted are *H. vittatum* and the hybrid plant; both of these are tetraploid with 44 somatic chromosomes. Since all the original hybrids appear to have been inter-fertile, it can probably be assumed that the other species would also show this count.

### ❀ *Lupin*

The basic species from which the garden Lupin has been developed is *Lupinus polyphyllus*. This is a stout perennial, with leaves that are covered with silky hairs on the underside, which may reach a height of 5 feet in the wild, although cultivated forms rarely exceed 3 feet. It is a native of western North America from California to the state of Washington. Its most frequent wild form is a purple-blue, but white, pale pink and bicolor forms have been recorded. Its improvement seems to date from the early years of this century when the Tree Lupin, *L. arboreus* was crossed with *L. polyphyllus* to give initially yellow strains and subsequently quite a rich crimson. Most Lupin species appear to be polyploid. *L. arboreus* with 40 chromosomes is apparently a tetraploid from a basic haploid count of 10, while *L. polyphyllus* with 48 chromosomes is probably a hexaploid from a basic count of 8. This polyploidy apparently allows the hybrids between species with different counts to continue to be fertile.

The appearance of the Lupin was transformed in 1937 when the strain developed by George Russell was exhibited in public for the first time. The late Mr Russell had been engaged in improving the plant for at least twenty-five years and had carried out the work by the most empirical of methods. Quite simply he collected Lupin species wherever he could and grew them all together. No deliberate pollination was undertaken and he relied entirely on insects for fertilizing the flowers. The only species that he admitted to using was 'an annual species sent from Germany'; presumably *L. hartwegii*, or possibly *L. pubescens*. These are both annual plants from Mexico with a variety of colour forms and they may well at some stage in their long career as garden plants have become hybridized, so that the race of annual Lupins is now a hybrid strain, in which also, according to some authors, *L. mutabilis* may have played a part. This is a more vigorous plant, also with various colour forms; the typical plant has white flowers with yellow on the standard, but cultivated forms are generally rose or purple. The plant is a native of South America, but has not apparently been found recently in a wild state. Any of these could have been the annual that was used by Russell.

Another species that may already have entered into the Lupin, before Russell started his work, is *L. nootkatensis*, which has become naturalized in one locality in Scotland. This is a littoral species from western North America with blue flowers variegated with red or yellow.

When the Russell Lupins were exhibited, one commentator thought that he saw in them traces of the following species: *LL. laxiflorus, lepidus, leucophyllus, mutabilis, nootkatensis* and *subcarnosus*. We have already mentioned *LL. mutabilis* and *nootkatensis*. Of the others *L. laxiflorus* is a perennial up to 2-foot 6-inches high with flowers that are either blue, pink or pale yellow and its presence would not be unlikely. On the other hand *L. lepidus* is a dwarf plant with violet flowers and it is difficult to imagine what it could have contributed. *L. leucophyllus* is a very hairy plant reaching as high as 3 feet with white, pink and bluish flowers. *L. subcarnosus* is a small-seeded annual with blue flowers with a white or yellow spot on the standard. This, again does not seem as if it would contribute very much, but it may have been present.

George Russell's treatment was simple in the extreme. As his seedlings flowered, anything that did not come up to his standard was destroyed and the strain is the result of rigorous selection of random crosses. It was apparently fifteen years before he started to get any spectacular results, but thereafter progress appears to have been fairly rapid. When Mr Russell was eventually persuaded to sell his stock to the firm of Messrs Baker, his plants so far surpassed any existing cultivars that they looked like a different flower altogether. The colours ranged from selfs in red, deep pink, orange and yellow, to bicolors in purple and gold, pink and amethyst, apricot and blue and many other combinations.

With a hybrid strain of this nature it is not possible to get cultivars to breed true from seed, so clonal propagation can only be done by cuttings. Thus the number of named plants is comparatively small, but the strain is kept in being and seeds still yield a wide variety of good colour forms.

The number of Lupin species is large, over three hundred, and in theory it should be possible to develop the genus still further. In practice many of the species are not easy in cultivation, so that the range may not be so large as the number of species suggests. But there is certainly scope for future hybridization, although the results must be regarded as uncertain. Many of the dwarf species, such as *L. lyallii,* have very attractive silvery leaves and since the garden Lupin has rather unsightly leaves, it would obviously be an advantage if they could be improved. The chromosome number of *L. lyallii* does not seem to have been recorded so it is not easy to say if such a cross is liable to succeed.

The shrubby species, *L. albifrons,* with a white standard and a blue or purple keel, also seems a species that might be used. It is not quite hardy in Great Britain, but could possibly be hybridized with *L. arboreus* to give a race of multi-coloured Tree Lupins. The pollen of *L. arboreus* has been used on *L. polyphyllus* to give yellow into the herbaceous species; the opposite cross might give more colours to the Tree Lupins. Lupins sound as though they should yet have surprises for us.

## ❀ *Mimulus*

In the *Floricultural Cabinet* for 1835, p. 119, there appears the following paragraph, which sums up the situation fairly thoroughly:

'Having recently been in York and its neighbourhood, we were much pleased to see several very strikingly handsome varieties of Mimuluses, which had been raised by cross impregnation from the *M. variegata* [*sic*], *roseus, luteus, Youngii, Smithii, bifrons* etc. The friends of floriculture who have thus been pleasingly employed, have been most agreeably compensated by the very handsome productions which have succeeded their labours. There is such a delightful anxiety connected with attempts to produce new varieties of plants by cross fertilization, that we strongly invite the attention of persons fond of floriculture to it. We are aware that in some instances original handsome species have been superseded by inferior flowers; but such should be cast away, and only equally handsome or superior be reserved. By attention to the kinds impregnated etc. a most decided improvement might be effected in the new kinds produced. As it respects the Mimuluses we have referred to above, there is a very great improvement in their beauty, beyond that of their parents. Lady Milner of Nun Appleton has been very successful in raising several pretty varieties; one is remarkably handsome, produced from seed of *M. variegatus*. The flower has a fine bold spot, as large as *M. Youngii*, of a fine deep striking purple colour, upon a light yellow ground. The plant is much more vigorous than its parent, and when in bloom a most beautiful object. Miss Nelson of York and Messrs Backhouse, nurserymen of York have each raised very distinct and strikingly handsome kinds, which deserve a place in every flower garden, very far exceeding any of the kinds previously grown.'

This is a reasoned defence of the hybridist's behaviour and interesting from the information it contains. Some of the nomenclature is now different. *M. variegatus* is still known by that name, but *M. roseus* is now *M. lewisii*. *M. youngii* (properly *youngeanus*) and *smithii*, were hybrids that had been previously raised and were probably crosses of *variegatus* and *luteus*. It seems probable that *MM. luteus* and *cupreus* may not have been distinguished in those days. Loudon's *Hortus Britannicus* mentions a cross between *M. cardinalis* and *M. roseus*, but this may well have proved sterile. We do not appear to have the chromosome number of *M. lewisii*, but *M. cardinalis* is a diploid with 16 chromosomes, while *M. luteus* and the modern hybrids are octuploid with 64 chromosomes. The count for *M. luteus* is a little suspect. The plant has been so long in cultivation that it has naturalized itself in some parts of the country and if one of these long-cultivated forms was the subject for the count, it may well be different from a genuinely wild form. *M. guttatus,* which is generally considered to be a form of *M. luteus,* was counted in 1940 and found to be hexaploid with 48 chromosomes. This may indicate that *M. guttatus* is a separate species or it may indicate that *M. luteus* is an aggregate of polyploids. The probability would seem to be that all these species have been intercrossed over so long a period that they now represent a hybrid swarm and the true species can only be found in their original localities. Darlington and Ammal give Chile as the home for *M. luteus,* but it is not found there, being replaced by the similar *M. cupreus,* so it is not quite clear what species was used. It looks rather as if *M. cupreus* might be an octuploid, while *M. luteus* is a hexaploid, but unless one knows the provenance of the plants used, all these counts must be accepted with some reserve. It is clear that the hybrid race is octuploid, which accounts for its variability and high fertility.

## ❀ *Montbretia*

This name is still attached to plants that are now classed as *Crocosmia.* There are only two species involved, both South African in origin. *C. aurea* was introduced in 1846. The most typical form is golden yellow and the flowers are borne on a scape about 2 feet high. In 1888 two more vigorous and larger-flowered varieties were introduced. Var. *imperialis* had larger flowers that were a brilliant orange-red and var. *maculata* had orange flowers with a dark blotch at the base of the petals. *C. aurea* is not always reliably hardy in Great Britain.

*C. pottsii,* named after its introducer, Mr Potts of Lasswade, entered into cultivation in 1877. It is a taller plant, with a scape reaching from 3 to 4 feet in height and flowers that are yellow inside and brick-red on the outside. It is somewhat hardier than *C. aurea.* The two species were crossed by Lemoine in 1880 and the resultant hybrid was given the name *Crocosmia* X *crocosmiiflora,* not, perhaps, one of the botanists' happiest efforts. Lemoine's hybrid was the plant that is popularly known as Montbretia, but subsequent crosses, taking in the larger varieties of *C. aurea* that were introduced after Lemoine's original hybrid, have given rise to a race of plants with large open flowers in colours ranging through all shades of yellow and orange. One of the principal breeders was Sydney Morris of Earlham Hall and the best cultivars are still known as Earlham Montbretias. Unfortunately these superior cultivars do not seem to have the stamina of the first indestructible Montbretia and the original stocks seem to be dying out. This points a moral for plant breeders. To recreate the large-flowered Montbretia it will be necessary to reintroduce the original species. They may still be in cultivation in Botanic Gardens, but they are certainly not in commerce. Of course they had only a limited popularity and it is probably on this account that they have been allowed to die out. They are probably due for a revival.

### ❀ *Nymphaea*

The first *Nymphaea* hybrids were reported in 1859 and were carried out by Herr Borsig at Moabit, near Berlin. He was working with greenhouse species and first crossed *N. rubra* with *N. lotus* and then crossed the resultant hybrids with *N. lotus* again.

The hardy hybrid water-lilies are almost entirely the creation of M. Marliac-Latour who worked at Temple-sur-Lot in the Department of Lot et Garonne in western France at the end of the last century. His hybrids were first described by Vilmorin in the *Revue Horticole* for 1891. He tended to be a bit reticent of the parentage of his crosses, which came out under the name *N.* X *marliacea*. However it is fairly certain that the species principally concerned are *NN. alba,* particularly its var. *rubra, odorata mexicana, tuberosa,* and *tetragona* (*pygmaea*).

*N. alba* is the common white water-lily of Europe and its var. *rubra* is found in Sweden. This has rosy-pink flowers and the plant known as 'Froebelli' is probably only a selected clone. A similar North American species is *N. tuberosa,* of which a pink form is known, but this is thought by Bailey* to be a hybrid between the pink form of *N. odorata,* which is found at Cape Cod, and *N. tuberosa.*

There would appear to be some question as to whether the two very small species *N. pygmaea* and *N. tetragona* are two names for one species. Most authorities assume that the epithets are synonymous, with *tetragona* having priority, but the late Amos Perry, who worked very extensively with Nymphaeas and other aquatics, was of the opinion that they were distinct. *N. tetragona* he claimed was purely Chinese, while *N. pygmaea* had a more extended range, and he claimed that the flowers were of different shapes. Both plants are small, growing in shallow water, with white flowers about 2 inches across. *N. mexicana* (*flava*) is, as its name implies, a native of Mexico and also of Florida. It is characterized by mottled leaves, a feature that it shares with *N. tetragona,* and yellow flowers, about 4 inches across, which are carried on their stems about four inches above the surface. It is not reliably hardy in northern districts and so it has been hybridized to give hardy yellow-flowered water-lilies. The presence of mottled leaves usually implies the presence of *N. mexicana* in the plant's parentage, although the dwarf plants of the Laydeckeri strain (named by M. Marliac-Latour after his foreman and son-in-law) which have *tetragona* in their parentage, will also show this mottling. The cultivar known as 'Helvola', the smallest water-lily known, which can be grown in a large goldfish bowl, is *N. tetragona* X *mexicana.* It has small mottled leaves and yellow flowers. M. Marliac-Latour tended to be reticent about the parentage of his hybrids, but it is known that some were *alba* X *odorata rosea, alba* var. *rubra* X *tetragona,* and *mexicana* X *tuberosa.* Other hybrids were raised by other growers, notably Dreer, who produced the splendid crimson 'James Brydon', but they were all working with the same species. One of the most popular crosses has been *alba* var. *rubra* X *mexicana* and the reciprocal

* In *Hortus,* 2.

cross. *N. tuberosa* obtained its name from the rootstock bearing tuber-like growths which detach themselves and form new plants. It is thus liable to spread with great rapidity, making its hybrids useful for large expanses of water, but unsuitable for small ponds. All the Nymphaeas are multiple polyploids. It has not, apparently, been decided whether the basic haploid number is 7 or 14, but if we assume that it is 14, *N. mexicana* is a tetraploid, while *odorata* and *tuberosa* are hexaploids. *N. alba* seems to exist in hexaploid and octoploid forms, while the form known as *N. candida* appears to have about 160 chromosomes; 168 sounds the most probable count, which would make it a dodecuploid.

On August 9, 1898, Marliac-Latour addressed the R.H.S. (*Journal*, Vol. 22, p. 287) and gave some details of his work and methods. In this he suggested that the best reds were crosses of the tropical *N. rubra* and its hybrids with *N. odorata rosea*. He used the last-named species as the seed parent and said that this gave the rich reds of the tropical species with the hardiness of *N. odorata*.

Although A. Niklitschek (*New Flora and Silva*, Vol. 4, p. 152) stated that all the hybrids were sterile, Marliac said that he grew some seed from his hybrids, but that they tended to segregate out. Apparently no $F^2$ generation was bred. He also gave some fairly detailed directions as to his method of hybridization, although few people seem to have been able to follow him.

'The work of hybridization,' he said, 'is more complicated as it is necessary to entirely cut away, at the very first moment of expansion, all the stamens . . . and on the second day to dust their stigmas with . . . pollen from those kinds chosen for the crossing of them.' After fertilization the ovary sinks and ripening takes place under water, but when the seeds ripen they float to the surface looking like small pearls. They must be watched for carefully and collected at once, as they only float for a short time, after which they fall to the bottom. If it is not desired to sow the seeds immediately, they are best kept in water and not allowed to dry out.

### ❀ *Pentstemon*

These plants are mainly hybrids between the two species *P. cobaea* and *P. hartwegii*. *P. cobaea* is a plant of central U.S.A. with 1-2 foot high stems and dull purple flowers, while *P. hartwegii* is a rather more vigorous plant from high ground in Mexico with scarlet flowers. It is possible that some of the putative hybrids are really selected forms of *P. hartwegii*, but the bulk of the plants are hybrids between the two species. *P. cobraca* arrived in 1825 and *P. hartwegii* ten years later and the hybrids were in existence in 1841. Originally there were named cultivars that were propagated by cuttings, but nowadays they are treated as annuals and grown from seed. The basic haploid number of *Pentstemon* is 8 and, of the few that have been counted, most are diploids with 16 chromosomes. *P. cobaea*, however, is an octuploid with 64 chromosomes and so, presumably, is *P. hartwegii*, although it does not seem to have been published. Since the hybrids are inter-fertile, it is to be presumed that they have the same chromosome count. There is another octuploid species *P. venustus*, with purple flowers, that could, one imagines, be used to vary the strain slightly; but probably the plant is not sufficiently popular to warrant any experiments.

### ❀ *Petunia*

The enormous range of colours and forms in the Petunia represents a triumph for the selectors, as only two species have been involved. These are *P. axillaris* (*nyctaginiflora*) and *P. violacea* (*integrifolia*). The first species has large, white, night-scented flowers; the flowers of the second species are smaller and the flowers are red or violet. *P. axillaris* was introduced to cultivation in this country in 1823, while *P. violacea* arrived in 1831. Hybrids had certainly flowered in cultivation by 1837 and the hybrid race has been continuously interbred ever since. Doubtless when both species and hybrids were available, the larger flowered cvs. had a double dose of *P. axillaris*.

During its century of cultivation, polyploids have arisen –

triploid, tetraploid and pentaploid forms have been recorded. In the process of breeding and selection, forms with double flowers and with fimbriated edges have arisen, besides a large number of different colour forms from white to scarlet and including violet shades. Double forms can be perpetuated by cuttings, but seeds from a cross with a double flower as one parent will throw a good proportion of doubles. The plant is treated as an annual, but *P. violacea* is perennial in its native Argentine and in the tropics the plants are not necessarily annual.

During the early nineteenth century the plant was cultivated in a manner which now seems extraordinary, as the following extract from the *Floricultural Cabinet* for 1841 (Vol. IX, p. 109) will show.

'Not having seen in any number of your CABINET a method of growing the Petunia as specimens on lawns, I beg to send you my plan of cultivating it. . . . In the beginning of February I take plants that were struck the previous autumn and had been potted into 48 sized pots. . . . I then pot them into 32s, filled with equal parts of loam from an old melon bed and leaf mould, with a little sand, which when filled with roots I shift them into 24s and lastly into 16s [9½-inch diameter].

'After I commence potting in February I place the plants in . . . 60 degrees of heat. If I find the plants not as sufficiently strong as I could wish, I pinch off the tops of the principal shoots which soon induces the strength required; and when they begin to make shoots of some length I commence training them to sticks, always giving plenty of room and this I continue till the end of May, at which time I plant them in the lawn.

'After choosing the place I intend it to grow, I remove the turf to make a hole sufficient to take two barrows full of the compost. . . . When the plant has been in this situation long enough to have made shoots 3 or 4 inches long, I drive in four stumps about 6 inches from the centre forming a square, having 2 feet above the surface; on these I fix a wooden hoop 3 feet in diameter. . . . To this I train the plant which soon covers it. After it has made shoots 6 inches long over this first hoop, I lay on another 5 feet in diameter.

'By the above mode of cultivation I have grown a plant of Petunia *twenty-one feet in circumference*, forming a complete table of beautiful rosy purple blossoms and much admired by all who have seen it.' The final statement sounds very convincing.

Although some forms of *P. violacea* are self incompatible, by and large Petunias can be self-pollinated, so that it is possible to fix the various colour forms. At the present day a number of $F^1$ hybrids are offered by seedsmen. Since they are working from a strain that has long been hybridized, it is not clear what particular advantages are to be gained from this hand pollination. Indeed the phrase $F^1$ hybrids is incorrect in this context. What we have is the result of hand pollination between two selected cultivars. There would probably be a greater uniformity in the resultant plants and it is claimed that they are markedly more floriferous. There seems no good reason why this should be so.

## ❀ *Phlox*

According to nineteenth-century writers the garden plants have been derived from *P. paniculata* (*decussata*) principally, with a certain amount of hybridization with *P. maculata* and *P. carolina*. This last-named flowers in late spring and crossed with the later-flowering species, it gave rise to forms that were less robust, but flowered earlier. Both *P. carolina* and *P. maculata* have hairy stems, whereas those of *P. paniculata* are nearly glabrous, a characteristic that is found in the garden race. The presence of *P. maculata* in the garden strain is by no means proved and it well may be that the later-flowering plants are simply selected forms of *P. paniculata*, while the earlier flowering plants are hybrids with *P. carolina*. The typical wild form of *P. paniculata* has violet-purple flowers, while white and pale-pink forms of *P. carolina* are found, as well as those of a purple hue.

### ֍ *Schizanthus*

Properly speaking Schizanthus should not be in this chapter as many of the plants offered under this name are selected forms of *S. pinnatus*. *S.* X *wisetonensis* is known to be a hybrid of *S. pinnatus* and *S. grahami* and it is probable that the plants known as Schizanthus nowadays are a hybrid mixture of these two species and *S. retusus*. The species are not easy to distinguish in any case. *S. retusus* is rather more compact and may be the predominant parent in the strain known as *compactus*. The species are all characterized by a yellow blotch on the upper petal, but this has been bred out in some strains and self-coloured flowers have resulted.

### ֍ *Streptocarpus*

The history of Streptocarpus hybrids dates from 1884, when seed of the brick-red *S. dunnii* were received at Kew. This flowered in 1886 and the plants were crossed with the blue *S. rexii* and the white *S. parviflorus*. The resultant hybrids were intercrossed and back-crossed and a range of colour forms resulted. About 1912 the emergence of a recessive gene added pink and salmon colours to the range. All the species are diploids with 32 chromosomes, but a tetraploid form, 'Merton Giant', has been raised in recent times. *S. rexii* is still occasionally seen, but the prime instigator of the hybrid race, *S. dunnii*, seems to be no longer in cultivation.

In 1859 *S. gardenii* had been crossed with *S. polyanthus* and in 1875 *S. rexii* and *S. saundersii* had been crossed, but neither of these hybrids appear to have been used subsequently. Two other *dunnii* crosses are recorded. 'Dyeri' was *dunnii* X *wendlandii* and '*watsonii*' was *dunnii* X *parviflorus*. Other species have been bred into the original *dunnii* X *rexii* cross. 'Achimeniflorus' was X *hybridus* X *polyanthus*, 'Banksii' was X *hybridus* X *wendlandii*, while 'Gratus' was *dunnii* X X *hybridus*. It may well be that the modern strain contains all these species, which are in various shades of mauve and violet.

*S. wendlandii* has purplish leaves and might still prove advantageous since this character has not appeared in the hybrid strain. Another species that might be employed is *S. denticulatus*, which has more flowers to the inflorescence than most species. It was used once in the hybrid 'Taylori', which has the involved parentage of a white form of the hybrid X *hybridus* X *polyanthus* crossed with *S. denticulatus*. This sounds a very promising cross and one would like to know what has become of it.

All the hybrid Streptocarpus come from the stemless group of the genus. The caulescent group, which has a chromosome count of 30, does not appear to have been much cultivated at any time and is only moderately decorative. Owing to the different chromosome count, hybridization between the two groups is not possible.

### ❀ *Verbena*

The bedding Verbena is a hybrid to which four species have contributed. The first hybrid was raised in 1837, and subsequent hybrids were later all interbred to give the strain that is now known as the bedding Verbena. The four species are *VV. platensis* (*teucrioides*), *peruviana* (*melindres*), *phlogiflora* (*tweediana*) and *incisa*. It will be appreciated that the taxonomy is rather baffling. During the 1870s the strain was greatly improved by Henry Eckford and it is to him that the modern strain with a white 'auricula' eye is due.

*Verbena platensis* is a sprawling perennial with a habit of rooting at the nodes. It has a rather elongated inflorescence composed of many white or pale pink flowers. *V. peruviana* is sometimes seen nowadays under the name *V. chamaedrifolia* or *V. chamaedryoides*. It is a prostrate plant with brilliant scarlet flowers. *V. phlogiflora* is a native of southern Brazil and Uruguay. It is a sub-shrub, usually rather sprawling, but can have stems several feet tall. It appears in numerous colour forms including purple, lilac, blue and red. From its description it sounds as though it would be worth growing on its own account and it seems a pity that it is no longer in cultivation.

*V. incisa* is another scarlet species, usually with a white eye, and may reach a height of 2 feet. Like all the species involved, it comes from the southern half of South America.

It is thus possible to see what each species has contributed. The most valuable must have been *V. peruviana* which gave a dwarf procumbent habit. *V. incisa* seems only to have contributed the white eye. The blue and purple shades came from *V. phlogiflora* and the white and pastel shades from *V. platensis*. At the end of the last century when the plant was at its most popular, there were very many named cultivars that were regularly propagated by cuttings. These had to be overwintered under glass and, with the disappearance of many greenhouses, they have been discontinued and now Verbenas tend to be treated as half-hardy annuals and grown regularly from seed. Strains have been evolved that come true to name and colour. In fact the hybrid strain has become so stabilized that it behaves as though it were a species.

In 1926 the hybrid Verbena was further crossed with *V. tenuisecta* (*erinoides*). This is a prostrate species with purple flowers and very finely-dissected leaves, the segments of which are thread-like. The resultant plants *Verbena* X *teasii* had very deeply-incised leaves, which were not so thread-like as in *V. tenuisecta*, and rather small flowers in great numbers in varying colours of blue, purple, pink and white. The scarlet does not seem to have appeared in this strain. I have no knowledge as to whether this new race of hybrids still exists. I have not seen it offered in any seed-list that I have consulted. If it has been lost, one would think that it was a cross worth repeating, but it may still exist in warmer climates.

The disappearing species are not confined to herbaceous plants; they also include the original species of two well-known shrubs, deciduous Azaleas and South African Heaths.

### Deciduous Azaleas

There are a very large number of Rhododendron hybrids in existence and their number is constantly increasing, but their creation is outside the scope of this book. Their parentage is well set out in the *International Rhododendron Register* (R.H.S., 1958) and in Part Two of the *Rhododendron Handbook* (R.H.S., 1964). The majority of these hybrids are named plants that are used as clones and propagated vegetatively. The exception lies in the various races of deciduous Azaleas.

The original races were known as Ghent Azaleas and Mollis Azaleas. The Ghent Azaleas appeared first and their development is attributed primarily to P. Mortier, by profession a baker, but an amateur horticulturalist. The hybrid that still bears his name is a cross between *Rhododendron calendulaceum* and *R. nudiflorum*. Both these species, like all the others used in the first Ghent hybrids, are of North American origin. The exception is the Eastern European and western Asian *R. luteum*.

*R. calendulaceum*, introduced to cultivation in 1804, is a striking shrub with flowers of a brilliant orange, tending sometimes towards scarlet and sometimes towards yellow. It is a very imposing plant. *R. nudiflorum* is a smaller plant, usually flowering slightly earlier than *R. calendulaceum*, with typically pale pink flowers. The other species used in the original Ghent hybrids are *speciosum, viscosum, roseum* and *luteum*. *R. speciosum* is not very dissimilar to *calendulaceum*, but tends to be more scarlet. *R. viscosum*, the Swamp Honeysuckle, is a rather small shrub with fragrant white flowers. It is much later flowering than the other species and so is valuable in prolonging the flowering season. *R. luteum* (*flavum*) is the popular yellow Azalea from the Black Sea region. It is markedly fragrant and the leaves colour well in the autumn.

The Mollis hybrids are derived from two Asiatic species, *R. japonicum* and *R. molle*. *R. japonicum* is a robust shrub, which, confusingly, was known originally as *R. molle*. It has large flowers ranging in colour from orange or salmon-red to brick-red and has a conspicuous orange blotch. *R. molle,* which adds to the confusion by being referred to frequently as *R. sinense,* is far less sturdy in Northern Europe and has yellow or pale orange flowers with a greenish blotch. The hybrids had the robustness of the *japonicum* parent and a good range of brilliant colours.

In the meantime British hybridists were at work and added the eastern species, particularly *R. molle,* to the Ghent hybrids and to the American species. Thus 'Aurea Grandiflora' is *luteum* X *molle* and 'Altaclarense' was *molle* X *viscosum*; another plant of the same parentage was 'Viscosepalum'. The two Asiatic species had far larger flowers than the American ones, but lacked any perfume.

Ghent and Mollis Azaleas were united in the Knaphill and Exbury strains. Moreover these added to the number of species with some other American azaleas and most notably *R. occidentale*. Unlike most of the other species in the *luteum* subseries, the flowers and leaves of *R. occidentale* tend to appear together; most of the others produce their flowers before the leaves unfurl. *R. occidentale* has white flowers, often flushed with pink, with a pale yellow blotch and seems to have been quite outstanding as a parent. One of the earliest Waterer hybrids was 'albicans' (*molle* X *occidentale*). The Belgian nurseryman Anthony Koster also used *R. occidentale,* crossing the Mollis hybrids with it and producing many fine plants.

Another species which it is conjectured may have played some part in the Knaphill hybrids is *R. arborescens,* a late-flowering white-flowered American species. More recently growers in the U.S.A. have added other native species to the number already employed. There can now be few of the *luteum* subseries that have not gone into the strain of hybrid deciduous Azaleas.

The original Ghent hybrids started to appear in 1829. The first Mollis hybrids were released in 1873. 'Albicans' was exhibited in 1894 and the Koster Occidentale hybrids appeared in 1901. The Knaphill strain appears to have been developed over a very long period from 1860 to 1924 and it is only comparatively recently that they have been widely distributed.

The history of the Azalea has been very clearly described in *Azaleas* by Frederick Street (London, 1859) to which this section is heavily indebted.

### *Erica*

We end this chapter with a sad cautionary tale. In the case of the South African Heaths, not only the species, but also the hybrids, have, with one exception, disappeared from cultivation.

During the latter half of the nineteenth century there was a great vogue for these Heaths, which could be got into flower in all months of the year. They were showy plants, requiring little heat. Indeed considerable attention had to be paid to prevent them from becoming overheated or from becoming excessively dry and presumably their disappearance was caused by the 1914-18 war, when insufficient man-power was available for their upkeep. Several nurserymen specialized in these plants and Messrs Rollisson of Tooting were particularly renowned for them. The *Floral World* for 1876, pp. 250-51, gives a list of no less than 143 species and hybrids that would give a display throughout the year. In point of fact the list is longer, as various colour-forms are given separate entries. Of all these plants only two are seen nowadays, both grown for the Christmas trade. *E. hyemalis* is a mystery. The plant was first noticed about 1845. It has never been found growing wild, but there seem no species that could give this result as a hybrid. The nearest species is *E. perspicua*, which flowers in the summer over here. It is, I suppose, possible that one seedling strain failed to acclimatize itself to the Northern Hemisphere and flowered here at the same time as it would in its native South Africa. Alternatively a late or an early flower may have been crossed with a genuine winter-flowering species and the resultant hybrid was *hyemalis*. The other winter-flowering Heath remaining in cultivation is the species *E. gracilis*, sometimes called *nivalis* for no very clear reason. With the resurgence of interest in greenhouse plants among small growers, there might be a future for these lost plants and they should, perhaps, be reimported and further hybrids bred.

It is a depressing thought that plants over which men have toiled for a long period should have vanished; not because they have been superseded by superior plants, but because the desire to grow them has vanished altogether.

CHAPTER XIV

# Variation by selection

In a sense all the plants that we have been considering are the result of selection. However, the majority of those discussed so far have been the result of deliberate hybridization. In this chapter we shall consider species that are sufficiently variable in their own right.

Some species are naturally more variable than others, but for the full potentialities of such variation to become apparent in horticulture certain features will be found to be constant. Firstly the plants must be sufficiently popular for large quantities to be grown each year. Secondly the plants must be those that are either annual or that flower reasonably soon after the seed has germinated. Thirdly it is a definite help, although not essential, for the plants to be self-fertile. Now we can find exceptions to all these conditions, although very few for the first one and if there is an exception here, it is probably owing to one man's interest having been aroused. One thinks of the Reverend Wilks and the Shirley Poppies. If he had not had his interest aroused by the one wild poppy with a white edge to the petals, it is doubtful if the strain would have originated. Even so, it will be recalled, he had tō grow on all the resultant seedlings and cull the typical scarlet flowers in the succeeding generations. The case of the Shirley Poppy is also interesting as demonstrating that variability may exist without being in any way apparent. We have all seen fields of poppies, in which every flower appears to be identical; many plants throw white flowers from time to time, but how often is a white poppy seen? One would have thought that *Papaver rhoeas* was a very stable species, with practically no variation, so far as flower colour was concerned. Yet if we look at some of the older gardening books, we see that variation had been noted. Sweet's *Hortus Britannicus* notes a white, a flesh-coloured, a white-edged and a scarlet-edged form and these were all depicted in Weinmann's *Phytanthoza Iconographica* of 1737. The suggestion of variability was there for anyone to develop. The same cannot be said of the perennial *P. orientale* which was, however, developed into numerous shades ranging through salmon to white by the late Amos Perry. However, there was a tradition of variability in the genus; both the Iceland Poppy (*P. nudicaule*) and the Opium Poppy (*P. somniferum*) had long been known for their various colour forms and, in the case of the Opium Poppy, doubleness was an early phenomenon. This was regarded as attractive and so nurserymen were encouraged to persist in breeding them, with the result that the Opium Poppy soon became a popular annual.

The principal exception to the second condition is to be found in the cultivars of *Paeonia lactiflora*, which require some five years between sowing the seed and seeing the first bloom. Moreover once the breeder has produced a flower that he considers to be first rate, the plant can only be propagated by division of the tubers, and increase is therefore very slow. Of course in a commercial establishment fresh batches of seed come into bloom each year, so that, once the initial five-year period is over, there is a continual appearance of fresh cultivars. In spite of that, it is probable that many potentially valuable crosses may be lost, because of the time and space it would take to grow on a second generation. If the primary cross is not successful, it will probably be scrapped. It is not commercially desirable to grow on a second generation, which may prove equally disappointing. If this is the case with slow-growing plants like Paeonies, it is even more the case with trees and large shrubs. Rhododendrons are excessively variable in many cases, but it is obviously better to perpetuate a good clone than to grow hundreds of seedlings in the hopes of finding perhaps one improved form.

With annuals and perennials that will flower in the second

year, the case is very different. The flowers are quickly seen with the result that promising lines can be followed up and unpromising ones can be equally speedily rejected. Three or five generations can be bred in the same time as one generation of Paeonies and progress is thus much more rapid.

Self-fertility helps one to 'fix' a colour more rapidly. When the blue Primrose was bred, it would have been of little use if only one plant was the right colour. Primroses have two forms, known as pin and thrum; the first has a long style, while the second has a very short one and fertilization is only possible between the two. A 'pin' must be fertilized by a 'thrum' or vice-versa. This device, known as heterostylism, is effective for insuring cross-pollination, which may be genetically advantageous, but is irritating for the plant breeder. There is a popular white crucifer, much used for bedding, called *Alyssum maritimum.* A few years ago one violet-coloured plant appeared in a seed-ground. Since Alyssum is self-compatible, it was possible to isolate this coloured strain at once. Had it not been, the crossing would have had to be carried on for a few generations before the violet strain could have been fixed. Self-compatibility is not essential but it is a help. The aim of the breeder is very often to fix a certain colour form, and, if this is due to a single genetic factor, it can be fixed in the second generation.

Plants that are exclusively self-pollinated would, one imagines, show very little variation, if any, but the history of the Sweet Pea shows that some variation will occur even in these cases. Presumably during the cross-over in meiosis there is sufficient redistribution of the chromosomes to cause certain recessive genes to emerge. The wild plant was received from Francesco Cupani by Dr Uvedale of Enfield in 1699. This had a reddish-purple flower. A white form was recorded in 1718. In 1731 or 1737 a pink form was observed. This was given the name 'Painted Lady' which suggests that it might have been a red bicolor. In 1793 a so-called 'black' flower (presumably a dark maroon) and a scarlet, which was probably a brilliant red, were recorded. The cultivar 'Painted Lady' was popularly supposed to have come from Ceylon. Unless this was a publicity stunt by a nurseryman, it is hard to explain this very misleading statement, which has been persisted in until quite recently. We still sometimes find books in which it is stated that the Sweet Pea is a cross between the purple *Lathyrus odoratus* and a pink plant from Ceylon. This is pure fiction; there is no form of *L. odoratus* found in Ceylon. Picotee forms were known in 1860, so that most of the colours had already emerged before controlled breeding began about 1880. This controlled breeding gave larger flowers and some different forms. The dwarf 'Cupid' strain emerged in 1893, first in California and nearly simultaneously in England, France and Germany. All seem to have arisen as a mutant from the cv. 'Emily Henderson'. In the same way the Spencer type Sweet Pea with a waved standard and an open keel appeared spontaneously in four different localities as a 'sport' of 'Prima Donna'. Once these forms had appeared it was easy to perpetuate them by allowing the plants to seed naturally.

From the plant breeder's point of view the ideal is a variable plant that is capable of cross- and self-pollination. Some plants are so variable that the varieties have sometimes been taken for different species. This seems to have been the case of the popular Gloxinia, which has been bred, probably entirely, from various forms of *Sinningia speciosa*. It is possible that *S. guttata* and *S. reginae* have played some part, but by not means certain. *S. speciosa* had for some time been a favourite greenhouse plant, but in spite of the handsome colours, the fact that the flowers all drooped downwards meant that their true beauty was hidden. In 1845 (according to George Gordon in *The Floral World*, 1876, p. 98; the R.H.S. *Dictionary* says in 1867*) John Fyfe, the gardener at Rothesay House, raised a plant, supposedly from a cultivar called 'maxima', which held its flowers erect. This was called *Gloxinia fyfiana*. When more richly coloured cultivars were crossed with 'Fyfiana' the resultant plants preserved the upright-flowering habit and the modern strain was developed. As *Sinningia speciosa* is so variable, the

* 'About 1860' in the second edition of the *Dictionary*.

early botanists had tended to give each form specific rank and it was thought that the greenhouse Gloxinia was the result of hybridization. Although we cannot be sure that no other species have been involved (the *Gesneriaceae* cross with great readiness and intergeneric hybrids are frequent) it seems probable that the modern Gloxinia is entirely a result of selection.

One of the most remarkable examples of random selection over a very long period is the case of the Auricula; a plant that has been cultivated for nearly four hundred years. It would seem probable that the original plant may well have been the natural hybrid *P.* X *pubescens,* between *P. auricula* and *P. rubra.* This theory, first put forward by Anton Kerner in 1875, has not been generally accepted. The most that can be said with any certainty is that the Auricula is a hybrid and that *P. auricula,* in one of its many forms, was probably one of the parents. By the end of the sixteenth century several colour forms were known. Clusius speaks of a plant he calls 'auricula ursi II' which is thought to be *P.* X *pubescens.* In 1596 Gerarde was growing a yellow, a purple, a parti-coloured and several reds. In 1629 Parkinson listed twenty cultivars including one white, seven yellows, one violet, two purple (one of which sounds like *P. rubra*), five red or pink. The remainder are not easily identifiable; one had tan-coloured flowers and one had 'fair yellow dasht about the edges with purple'. The tan-coloured one had a white eye and farina on the leaves. Around 1660 both striped and double flowers were bred. The striped flowers look rather as though they may have had a virus infection and it may well be that they were allowed to die out, because they were infected and became weak. The striped flowers had farina both on the leaves and on the 'eye' of the flower, but this seems to have been only a light dusting. However, early in the next century the forms known as 'Bizarres' and 'Painted Ladies' had a heavy covering of farina and were the forerunners of the Show Auriculas. Around 1750, green appeared among the colours in the Auricula. This seems to have been a type of periclinal chimera in which the green of the calyx is reproduced in part of the petals, most frequently at the edge, but sometimes in the centre.

Shortly after the introduction of the green Auricula, various rules were laid down defining what was regarded as the most desirable type. It seemed generally agreed that the ideal Auricula should consist of four circles; the first and smallest is the tube. In a show flower this must always be thrum-eyed; a pin-eyed flower, whatever other desirable qualities it might possess, would lose points in a show. The second circle is the eye (referred to as the bottom in seventeenth- and eighteenth-century literature on the subject). In the Show Auricula this should have a heavy covering of farina, or paste as the growers call it. The eye is surrounded by a coloured circle, which is given the name 'ground'. Thus you have red-ground, violet-ground, green-ground, yellow-ground Auriculas. In the so-called Selfs, this colour will extend to the edges, but in the Fancy or Edged Auriculas there will be a fourth circle, which will extend around the edge of the petals. It is obvious that such meticulously grown flowers could only appeal to enthusiasts and for others there was evolved the so-called Alpine Auricula, which can have an eye of any colour, which is surrounded by a circle of the ground colour. The Alpine Auricula should not have any farina at all. For the general public there are Border Auriculas, which are similar to the Alpines, but may or may not have farina. It is this class of auriculas that is chiefly grown today. The Show Auricula is a private hobby and may well die out through lack of interest, although at least one nursery still grows them, so presumably there are still sufficient enthusiasts to make it commercially profitable.

From the historic point of view, the Auricula is of the greatest interest. There are no other European garden flowers with so long a record of continuous breeding and cultivation, with the exception of the Carnation. There are however some which have as long a history, but which do not seem to have been so developed. This may be because they were not florists' flowers and did not qualify for the show bench.

Both the Wallflower and Stock have been grown for a long

time, for their perfume. The Wallflower is grown chiefly as a single flower, although doubles are known; with the Stock it is the double flowers that are preferred. In the case of Stocks the doubling is absolute, both stamens and styles are petaloid and the flower is completely sterile. In the Wallflower this is not the case and it is possible to get seed that will throw almost all double flowers. The Wallflower has been known in England in a number of colour forms since 1573, but recently the range of colours has been much enlarged, presumably through controlled pollinations and the pink and purple shades have been isolated and extended.

The case of the Stock, *Matthiola incana,* is more interesting. The plant is known to have been grown in gardens since 1536. Two colour forms only were known, the purple of the wild plant and the white. Six years later a red form was noted. Double flowers first appeared around 1563. The pale yellow-flowered form was recorded in 1629. The unbranched, single-spiked Brompton Stock appears to be represented in the *Hortus Eystettensis* of 1612, but other authorities give 1660 for its appearance. The annual ten-week Stock is said to have been introduced to England from South Europe in 1731. The wild plant has the leaves covered with fine grey hairs, but by 1762 a less hoary plant had been bred and completely glabrous plants have been known since 1861. Rather improbably, doubling appears to be a simple Mendelian recessive and from normal seed one would expect the usual ratio of three singles to one double. However, since 1629, it has been recognized that there is one strain, known as the eversporting, which would produce considerably more doubles than the other. I have already given the probable explanation of this phenomenon (see Chapter II, p. 46). Following on this, the Danish nurserymen Hansen have been able to breed a strain in which the cotyledons (seed-leaves) of the single and double plants differ in colour, so that it has become possible to grow plants that are all double; the singles are discarded in the seedling stage. The perpetuation of the eversporting strain is an example of how horticulturists can preserve strains that would, in the normal evolution of the wild state, become extinct. From the evolutionary point of view a strain of which fifty per cent is sterile can have little survival value. It is probable that the strain only arose in cultivation and it has been preserved for at least three hundred years.

One of the most surprising examples of variation due entirely to selection is to be found in *Primula malacoides*. The plant was only introduced from China by Forrest in 1908. It was a pale mauve, small-flowered, rather dainty plant, which was grown for its winter flowering. Under cultivation, changes were soon observed. There was a certain amount of colour variation, although nothing more than one would ordinarily expect. In 1913 a double-flowered form arose and by the following year the size of the flowers had nearly doubled in diameter. Around the latter 1920s tetraploids appeared spontaneously and, although the first of these were rather shy flowerers, a few years' selection gave a tetraploid race of large flowers, which were produced in quantity. At the same time selection broadened the colour range. An albino form could have been expected and did not take long to appear, but the development of a carmine or magenta flower took longer. Now we have a complete spectrum from white, through lavender, mauve and purple to pink and carmine. Double flowers are not uncommon and the plant is a tribute to the activities of the breeder

*Primula obconica* has been longer in cultivation; it was introduced in 1880, but it does not show the variability of *P. malacoides,* and the colour range is more restricted. It has been hybridized with *P. malacoides,* but the results do not seem to have been very gratifying and the cross has not been persisted with. Although *Primula* spp. hybridize with great readiness, there seems to be no record of a hybrid giving rise to an improved garden form. Perhaps we should except the Polyanthus from this, if it is really derived from a cross between the Primrose (*P. vulgaris*) and the Cowslip (*P. veris*). It has however been argued that the Polyanthus is a longer stalked variant of *P. vulgaris*; that all Primroses are really growing in an umbel, but that in the common form the peduncle never emerges above ground. My own opinion, for what that is worth, is that the hybrid origin seems the more probable. One quite often sees natural hybrids of the two species in the wild and they do somewhat resemble the Polyanthus. The leaves of the Polyanthus seem to be intermediate in character between the two species. It is probable that this hybrid may have been repeatedly back-crossed to *P. vulgaris,* which would give larger flowers and longer pedicels on the umbel. For a long time the Polyanthus was a florists' flower and the 'laced' flowers, those whose petals had a yellow edge, were regarded as the most desirable. In other ways the desirable qualities were similar to those required in the Auricula and, as in that plant, only thrum-eyed flowers were exhibited. The early fanciers hand-pollinated their flowers, using a pin-eyed plant for the seed parent and a thrum-eyed one for the pollen. Whether this was because it was simplest, or whether there were more esoteric reasons for this process, is not clear. Fortunately the Polyanthus is easier to grow than the Auricula and the fact that it has ceased to be a florists' flower has not banished it from the garden. Some twenty years ago, seedsmen offered strains of coloured Cowslips, which were presumably the result of back-crossing Polyanthus on *P. veris.* These plants seem to have been discontinued and, indeed, there seemed little point in them, as they were less decorative than the Polyanthus. *Primula vulgaris* itself has many colour forms in its extensive range in the wild. *P. vulgaris* ssp. *sibthorpii* from Greece and the eastern Mediterranean regions has pink or red flowers, while ssp. *heterochroma* from Persia has a white and a blue form recorded. This latter is rare. The various colour forms have been known for long in gardens and have been bred together to give a wide spectrum of colours. They have also been crossed with the magenta *P. juliae* to give rise to a race known either as 'juliana' or 'pruhoniciana'. These plants tend to be rather strident in colour and have not the popularity of either coloured Primroses or Polyanthus.

A florists' flower that has descended in status and become a popular biennial is the Hollyhock. Of obscure origin, though probably Asiatic, *Althaea rosea* was introduced to Europe in 1573 but may have been in cultivation in Persia before then. Plants were bred with care, particularly by the nurseryman Chater, and were propagated by cuttings and division. The great increase in the onset of the fungus disease Hollyhock Rust and the inability to find any control measures meant that the perpetuation of cultivars became well-nigh impossible. It was always the older and more established plant that became infected and so the supply of named Hollyhocks had to stop. It was still possible to grow the plant as a biennial, as infection seldom occurred until the plant was just coming into flower. Plant breeders are now working on a rust-free strain and once this is achieved the cultivars of Hollyhocks may reappear. Nearly a century after his first cultivars Chater's strain is still grown.

The Snapdragon, *Antirrhinum majus*, almost suffered the same fate, as again a rust disease suddenly became epidemic. Fortunately this happened more recently and breeders have been able to evolve rust-resistant strains. The wild form of this plant is purple and at the start of the last century the only forms known were a scarlet, a pink, a white and a red and white bicolor. There was also a double form recorded, which seems to have disappeared. As a result of selection a number of other colours, notably yellow and orange, have been isolated and there are dwarf, medium and tall strains. A large flowered tetraploid strain has also been bred and the latest development is the 'Pentstemon-flowered' strain, in which the flowers have lost their characteristic shape and have been developed with tubular flowers that resemble the Pentstemon. Double flowers do occur occasionally in seed mixtures, but the double Antirrhinum is not popular and is not bred intentionally. Some years ago hybrids were bred between *A. majus* and the dwarf Pyrenaean *A. molle* to give a brightly coloured dwarf creeping plant for the rock garden. These do not seem to have met with any approval and are no longer offered, in this country at any rate. The range of colours and dimensions in *Antirrhinum majus* is one of the most striking testimonies of the value of systematic selection.

Only a little less spectacular is the range of *Senecio cruentus* and *Zinnia elegans*. The Cineraria, *Senecio cruentus*, is a native of the Canary Islands and bears a panicle of purple starry flowers. The 'Stellata' strain of Cinerarias is similar in habit to the wild plant, but the large-flowered strain is very different. It does not seem to have taken the breeders long to isolate blue and white strains, in addition to varying shades of purple, and many of the large-flowered strain have white eyes. More recently a coppery-scarlet and a deep blood-red have been added to the colour range. Although other species have been occasionally hybridized with *S. cruentus*, there seems no reason to believe that the garden Cineraria is anything except selected forms of the one species. In the 1870s double-flowered cultivars were raised, but these had to be propagated vegetatively and appear to have died out. Probably they could be bred again if there was thought to be any interest in them. Since the Cineraria is not a very graceful plant and is chiefly required to make a bold mass of colour, double flowers would perhaps enhance its value, as the flowers would probably be longer-lasting and it might be well worth someone's while to try to breed this double strain again. At the time of their first breedings, it does not appear as if this doubleness could be relied upon from seed and so the plants had to be propagated by cuttings. This is obviously commercially disadvantageous and it may be that this is the reason for their disappearance.

*Zinnia elegans* is a naturally variable species and, like other Mexican plants, may have a history of Aztec cultivation before being introduced to western gardens. The native species appears to be usually either violet or scarlet, but many other colour forms occur occasionally and the present race contains every colour with the exception of blue. A greenish flower is the latest to have been bred and the flowers tend to be composed entirely, or nearly entirely, of ray florets and give a double appearance. There is a dwarf race, a race with quilled petals and now a race with striped petals. Another, dwarfer species, *Z. haageana*, which normally has rather harsh orange-scarlet flowers has also been selected to give a multicoloured race of medium-sized plants, which are more successful in wet summers in Great Britain. In this country excessive wet tends to cause Stem Rot, so that the plants fail to develop satisfactorily and *Z. haageana* seems less prone to this disease.

Another Mexican composite that has been enormously improved by selection is Tagetes. There are two species in cultivation, *T. patula,* the French Marigold and *T. erecta,* the African Marigold. They have been known in European cultivation since the end of the sixteenth century and the original variation in colour does not seem to have altered much over the course of years. The African Marigold has rather an offensive smell, which breeders are trying to eliminate and it is interesting to see that Miller had a sweet-scented African Marigold in the mid-eighteenth century. This must have been subsequently lost and has only recently been re-bred. Miller notes that 'These are all subjects to vary, so that unless the seeds are very carefully saved from the finest flowers, they are very apt to degenerate.' Another point that Miller makes is that if selection is carried out in the same garden for too long, there will be degeneration in spite of the care in selecting the best flowers; 'therefore those who are desirous to have these flowers in perfection, should exchange their seeds with some person of integrity at a distance where the soil is of a different nature, at least every other year.' This, at least, is no longer necessary. Miller had pale and deep yellow, light and dark orange African Marigolds and the only subsequent addition in colour is a pale cream verging on white. It would appear that the plant did degenerate after Miller's day and that the doubles disappeared for a period, being reintroduced as a great novelty in this century. During the latter half of the nineteenth century, annuals were employed principally for bedding and, for some reason, the Tagetes were not used for this. Thus they had to be selected *de novo* when they came back into fashion. The French Marigold, the smaller *T. patula,* seems to have been more constantly in cultivation, but it is a less spectacular plant than *T. erecta.*

It will have been noted that the majority of these plants that have been so signally improved by selection are annuals. The list could be extended with such plants as *Godetia, Clarkia, Calendula* and many others.

Perennials seem to show less variability or have not been worked on so much. We have already mentioned Amos Perry's success with the Oriental Poppy, but such feats appear to be rather rare. Most variable perennials seem to have a certain amount of hybridity in them, such as the Michaelmas Daisies, although in this particular case it is not easy to decide where hybridity begins and species selection ends. The various forms of *Erigeron speciosus* are probably selected clones from seminal variants and the same can be said of the cultivars of *Gaillardia aristata. Monarda didyma* shows a certain amount of variation, but perhaps the most varied of perennials is *Chrysanthemum coccineum,* the Pyrethrum. The flowers of this vary from white, through various pink shades to dark crimson, while single and semi-double flowers are known. Most of the other variable herbaceous perennials are hybrids. Trollius, for example, are the result of crosses between *T. europaeus* with *T. asiaticus* and *T. chinensis.* The Red Hot Pokers are crosses between *Kniphofia uvaria* and other species. The garden Verbascums are mainly of hybrid breeding; not a difficult thing as they hybridize in the wild with great readiness. Whether it is that perennials tend to be more stable or whether they have been less worked is not clear. Probably both explanations are valid. Sometimes, as in the case of *Papaver nudicaule,* the Iceland Poppy, there is considerable variability, but the various colour forms cannot be isolated so that they come true from seed; or rather only a certain number can. One can buy seeds of the yellow and orange forms and the white is fairly reliable from seed, but the pastel shades of pink and salmon have not been segregated to breed true. Certainly there is nothing among perennials to show the variation that has been obtained in *Antirrhinum majus.* In the case of shrubs and trees variability is probably not to be encouraged. There is the case of Rhododendron, which is very variable in some species and it is the best forms that are vegetatively propagated, but no one is sowing Rhododendron seed with the object of obtaining improved forms. There is a fair amount of deliberate hybridizing in this genus, but the time between sowing and flowering is too long for a system of selection among seminal variation of species to be worth while. It is probable that there are many herbaceous perennials and shrubs that would be capable of improvement and variation through selection, if it took anyone's fancy. We should have to have someone with the vision of Amos Perry and such figures are rare.

## CHAPTER XV

# The question mark

The reader would be justified in complaining that this book has already more than its full share of unsolved queries. I can only agree. The particular question mark with which this chapter is concerned is the one that hangs over the future.

We may take it as reasonable that the breeders will continue to produce new Roses, Chrysanthemums and Sweet Peas, but what new plants are now going to attract the breeder?

The most spectacular race of plants to appear in recent years has been the De Graaff Lilies. Hybridists had worked with Lilies for some time previous to the introduction of these outstanding plants, but on a comparatively reduced scale. A hybrid Lily, thought to be *L. concolor* X *L. dauricum*, was brought from Japan in 1830 by von Siebold and given the name 'maculatum atrosanguineum'. This plant appears to have been crossed with the European *L. croceum*, the Orange Lily, and the resultant plants can be found called either X *umbellatum* or, more correctly, X *hollandicum*. During the 1890s Mrs R. O. Backhouse crossed *L. hansonii* with various forms of *L. martagon* to produce the Backhouse Hybrids.

In 1932 the Bellingham Hybrid Lilies were released for commerce. They derive from three North American species *LL. pardalinum, humboldtii* var. *ocellatum and parryi.* These are all 'turkscap' Lilies and fairly vigorous, although the hybrids show remarkable heterosis and frequently send up stems exceeding 6 feet in height. Subsequently other North American Lilies have been added to the strain.

The Mid-Century Hybrids are the results of crossing *L.* X *hollandicum* with *L. tigrinum*, the Tiger Lily, and back-crossing this with a selected clone of *L.* X *hollandicum* with the cultivar name of 'Alice Wilson'. The resultant hybrids had upright cup-shaped flowers in varying shades.

The Mid-Century Hybrids have been crossed in their turn with the turkscap flowered *L. davidii* to give a tall plant with upright flowers. They have also been crossed with the dwarf *L. amabile,* using both the red and yellow forms. These give dwarfer plants and bring yellow into the strain. The Mid-Century Hybrids have also been crossed with the graceful pink *L. cernuum* to give the 'Harlequin' strain, which has a wide range of colours and which brings back the turkscap flower to the hybrids. It is interesting that Dr C. F. Patterson seems to have obtained similar results by crossing *L. tigrinum, L. davidii* and *L. cernuum* together. Apparently *LL. concolor* and *croceum* have not contributed much to the 'Harlequin' strain.

Aurelian Lilies, which are crosses between Lilies with trumpet-shaped flowers and the tall, turkscap, *L. henryi,* date from about 1900, when *L. brownii* (*leucanthum*) was the seed parent. The plant did not survive. In 1925 E. Debras used *L. sargentiae* as the seed parent and called the resultant plant *L. aurelianense,* after his native town of Orleans. Even more remarkable was 'T. A. Havemeyer' in which the seed parent was the rather tender *L. sulphureum.* It flowered for the first time in 1937. When Mr de Graaff started breeding Aurelians he used a large selection of trumpet Lilies, and, more importantly, made the complementary crosses, using *L. henryi* as the seed parent as well. One of the results of this vast programme was the wonderful golden trumpet Lily 'Golden Clarion'. Finally there are a number of hybrids between the various trumpet-flowered Lilies, known as the 'Olympic Hybrids'. These are crosses between *L. regale, L. sulphureum, L. sargentiae,* and *L. brownii.*

All these Olympic hybrids are impressive flowers, which unfortunately show a great tendency to virus infection and it is difficult to keep the best clones healthy. Plants raised from seed will not have virus initially, but are rather liable to acquire it before they reach the flowering stage. *L. sulphureum* is particularly susceptible and tends to impart this susceptibility to its offspring. It is this weak constitution that prevents the Olympic Lilies being as popular as their merits deserve. One wonders if it might be possible to use some robust species such as *L. martagon* with the trumpet Lilies to get a more resistant strain. Until this problem is overcome the future of hybrid lilies must be regarded as a little doubtful.

Apart from lilies, it is not easy to see any new garden flowers being created. If I were a plant breeder I think I would investigate the possibilities inherent in *Polemonium* spp. The best known of these is an old garden plant, known as Jacob's Ladder, with a spike of blue flowers. However there are also yellow and pink species and albinos are not uncommon. They come from a wide range of habitats: North America, Europe and the Himalayas, and are found in a variety of forms from dwarf alpines to tall herbaceous plants. As far as is known they are all diploids with 18 somatic chromosomes and so, in theory, should hybridize easily and produce a large number of different forms and colours. So far no one seems to have done anything about them and they sound as though there should be possibilities.

Presumably the objectives of the plant breeder are to produce plants with a long flowering season, a variety of colours, a robust constitution, growth that does not need support (that is in herbaceous plants) and flowers that are either long-lasting or that are very numerous. The hybrid Hemerocallis have flowers that survive for only 24 hours, but they are being constantly renewed. If the flowers last well in water, it is obviously an advantage, but a failure in this respect has not militated against the popularity of plants such as the Russell Lupins.

Nowadays the choice of plants suitable for development and the first stages of this development would seem to depend on the enthusiastic amateur, to whom all gardeners already owe so much. Plant breeding needs time and space and the commercial nurseryman has little of either. Time is the most important, although it is obvious that results are more liable to be obtained if large numbers of plants can be raised. The nurseryman, already hard-pressed to make his business profitable, cannot afford to speculate in what may eventually prove unsuccessful. The enthusiastic amateur can do his breeding as a hobby and, if he is successful, the commercial breeder will step in and exploit his success. This may seem unfair, but the amateur will be paid for the results of his hobby and the public is the ultimate gainer. As we have seen, that is the way that many garden flowers have come into circulation. The Narcissus is almost entirely the creation of amateurs; the modern Iris hybrids owe much of their splendour to the hobby of Sir Michael Foster; the Hemerocallis has been developed mainly by George Yeld and Dr Stout.

Sometimes the place of the amateur is taken by research stations. Although they deal mainly with fruit and vegetables, some horticultural work stems from these institutions. Such, for example, was the source of the Bellingham Hybrid Lilies already mentioned. Even in this case, however, the initial impulse had come from an amateur and it would seem that it is due to a particular vision that some people possess, which perceives the possibilities inherent in the species. Once this vision takes possession, the rest may follow. It will not be the fault of the amateur if it doesn't. His vision may have led him to consider crosses that are impracticable, but if his ideas are sound, he will probably sooner or later obtain results. Naturally they will not come up to his hopes; perfection is not found in plant breeding nor anywhere else, but they may well contribute new plants to the garden. This may not seem very exalted. I disagree. To give many people pleasure and to harm nobody seems to me one of the most beneficent of occupations and this, after all, is what the plant breeder does. For the commercial grower this may mean his own and other people's living, but for the amateur it is pure art for art's sake and if he seeks renown as well, why should we blame him? It is a harmless enough ambition.

# Bibliography

*The R.H.S. Dictionary of Gardening*. Edited by F. J. Chittenden and P. M. Synge. Oxford, Clarendon Press, 1951, 4 vols. Second edition, 1956; Supplement, 1965.

ALLWOOD, M. *Carnations, Pinks and all Dianthus*. 3rd edition, 1947.

ANDERSON, A. W. *How we got our Flowers*. London, Benn, 1956.

BAILEY, L. H. and BAILEY, E. Z. *Hortus*. New York, Macmillan, 1930.

BURBIDGE, F. W. *The Narcissus, its history and culture*. London, 1875.

CANDOLLE, A. DE. *Origine des Plantes cultivées*. Paris, 1883. Translated as *Origin of Cultivated Plants*. London, Routledge, 1884.

CHATÉ, E. *Le Canna*. Paris, *c.* 1869.

CRANE, M. B. and LAWRENCE, W. J. C. *Genetics of Garden Plants*. London, Macmillan, 1947. Sixth edition, 1952.

CLIFFORD, D. *Pelargoniums*. London, Blandford, 1958.

CUTHBERTSON, W. *Pansies, Violas and Violets*. London, Macmillan, 1898.

DARLINGTON, C. D. and AMMAL, E. K. JANAKI. *Chromosome Atlas of Cultivated Plants*. London, Allen and Unwin, 1945.

DEAN, R. (and others). *The Dahlia: its History and Cultivation*. London, Macmillan, 1897. Revised edition, 1903.

DON, GEORGE, THE YOUNGER. *A General System of Gardening and Botany . . . founded upon Miller's Gardening Dictionary*. London, 1832-8.

DYKES, W. R. *The Genus Iris*. Cambridge, University Press, 1913.

FARINI, G. A. *How to Grow Begonias*. London, Sampson Low, 1896.

FARRER, R. *On the Eaves of the World*. London, Edward Arnold, 1917.

FERNANDES, A. Papers in *The Daffodil and Tulip Year Book*. London, 1968.

FORTUNE, R. *Three Years Wandering in the Northern Provinces of China*. London, 1847.

GRIEVE, P. *A History of Variegated Zonal Pelargoniums*. London, 1868.

HAGEDOORN, A. L. *Plant Breeding*. London, Crosby Lockwood, 1950.

HALDANE, J. B. S. *See* Winton.

HALL, Sir Alfred Daniel. *The Genus Tulipa*. London, Royal Horticultural Society, 1940.

HERBERT, DEAN W. *Amaryllidaceae*. London, 1837.

—— *A History of the species of crocus*. London, 1847.

HIBBERD, SHIRLEY. *The Amateur's Rose Book*. London, 1874. Revised edition, 1894.

—— (editor). *The Floral World and Garden Guide*. London, 1858-1880.

HURST, C. C. Articles in the *Journal* of the Royal Horticultural Society, London, 1941. (*See also* THOMAS, GRAHAM.)

JACKMAN, G. and MOORE, T. *The Clematis as a Garden Flower*. London, 1873. Revised edition, 1877.

JEFFERSON-BROWNE, M. J. *The Daffodil: Its History, Varieties and Cultivation*. London, Faber, 1951.

LEACH, DAVID. *Rhododendrons of the World*. London, Allen & Unwin, 1962.

LI, H. L. *The Garden Flowers of China*. New York, Ronald Press, 1959.

L'OBEL, MATTHIAS DE. *Plantarum seu Stirpium historia*. Antwerp, 1576.

LOUDON, J. C. *Hortus Britannicus*. 1839.

MILLER, P. *The Gardener's Dictionary*. 6th ed. 1771.

MILLET, A. *Les Violettes*. Paris, 1898.

PARKINSON, J. *Paradisi in Sole Paradisus Terrestris*. 1629. Facsimile edition published by Methuen, London 1904.

PHILLIPS, G. A. *Delphiniums, their History and Cultivation*.

London, Butterworth, 1932.

PORCHER, FELIX. *Le Fuchsia: son histoire et sa culture.* 1848.

RANDOLPH, L. G. (editor). *Garden Irises.* New York, 1959.

REDOUTÉ, P. J. *Les Liliacées.* Paris, 1802-16.

—— *Les Roses par P. J. Redouté.* With text by C. A. Thory. Paris, 1817-24.

RIVERS, THOMAS. *Rose Amateur's Guide.* London, 1837. Eleventh edition, 1877.

STOUT, A. B. *Daylilies.* New York, Macmillan, 1934.

STREET, F. *Azaleas.* London, Cassell, 1959.

SWEET, ROBERT. *Geraniaeceae.* London, 1820-30.

—— *Hortus Britannicus.* London, 1826. Third edition, enlarged and edited by George Don, London, 1839.

SYME, J. T. B. *English Botany.* London, 1863-86.

THOMAS, GRAHAM. *Old Shrub Roses.* London, Phoenix, 1955. (Including reprints of papers by C. C. Hurst, first published in the R.H.S. *Journal* for 1941.)

—— *Shrub Roses of To-day.* London, Phoenix, 1962.

—— *Climbing Roses, Old and New.* London, Phoenix, 1965.

THUNBERG, C. P. *Flora Japonica.* Leipzig, 1784.

WEINMANN, J. W. *Phytanthoza Iconographia.* Ratisbon, 1737-1745.

WILLIAMS, W. *Genetical Principles and Plant Breeding.* Oxford, Blackwell, 1964.

WILLMOTT, ELLEN ANN. *The Genus Rosa.* London, John Murray, 1910-14.

WINTON, D. DE and HALDANE, J. B. S. 'The genetics of *Primula sinensis*' in the *Journal of Genetics,* 1933.

WOOD, W. P. *A Fuchsia Survey.* London, Williams & Norgate, 1950.

*Addisonia: coloured Illustrations and popular Descriptions of Plants.* New York, 1916-

*The Botanical Magazine* by W. Curtis. London, 1787-1800. Continued as *Curtis's Botanical Magazine.* London, 1801-1947.

*The Botanical Register* by Sydenham Edwards. London, 1815-1828. Continued as *Botanical Register or Ornamental Flower Garden* edited by John Lindley. London, 1829-47.
*The Floral World and Garden Guide*. Edited by Shirley Hibberd. London, 1858-1880.
*Flore des Serres et des Jardins de l'Europe*. Ghent, 1845-83.
*The Floricultural Cabinet and Florist's Magazine*. Conducted by J. Harrison. London and Sheffield, 1833-59.
Journal of the Royal Horticultural Society, London, 1866-
Journal of the British Iris Society. London.
*The New Flora and Silva*. London, 1928-40.
*La Revue Horticole*. Paris, 1842-

# Glossary

ALLELE, ALLELOMPORPH. A gene having two or more contrasting functions. That most usually manifested is termed the dominant gene, those that tend to be obscured are recessive.

APOMIXIS. The production of fertile seed without normal fertilization. Such plants are termed apomictic.

BUD SPORT. A mutation in a section of a plant. Very often a single flower or a branch will differ in appearance from the bulk of the plant. If the section of the plant where the sport appears is propagated vegetatively the mutation is perpetuated.

CHIMAERA. A hybrid state in which the cells of each of the species involved, though closely linked together, continue to retain their own characteristics. In a PERICLINAL CHIMAERA the cells of one species form a layer which completely covers the cells of the other species involved in the hybrid.

CLONE. A population derived by sterile (vegetative) propagation from a single plant. This is often necessary when some particularly desirable form occurs, either as the result of hybridity or by seminal variation.

CULTIVAR. Forms of plants that have arisen in the course of cultivation and are not known in the wild. This principle is sometimes breached, as when a wild form of particular attraction is used to produce a clone, which must be distinguished by a cultivar name. Cultivar is abbreviated cv.

DICOTYLEDON. Plants that produce two seed-leaves when the seeds germinate

DIPLOID. Having twice the basic (haploid) number of chromosomes in each nucleus.

DIOECIOUS. Bearing male and female flowers on separate plants.

DISK FLORET. The small, petalless florets in some members of the *Compositae*, such as the Daisy and Groundsel.

GIGANTISM. A condition in many cultivated plants where the cells are much larger than in their wild ancestors. As a result the plant is larger in all its parts. This occurs notably in most of our vegetables and also in most forms of the florist's Cyclamen. The opposite of this, which causes plants much smaller than normal is sometimes termed nanism.

GLABROUS. Devoid of hairs on the leaves or stems.

HAPLOID. The basic chromosome count for a genus or a species. The nucleus will contain multiples of the haploid number.

HETEROSIS. Hybrid vigour. Many first crosses either between species or between cultivars show considerably more vigour than either of the parents.

HEXAPLOID. Having six times the haploid number of chromosomes in each nucleus.

HETEROSTYLY. Bearing styles of different lengths. Adopted by many species to ensure cross fertilization. Most conspicuous in the pin and thrum flowers of *Primula* spp.

HETEROMORPHIC. Appearing in more than one shape.

IMBRICATE. With the leaves or petals overlapping each other like tiles on a roof.

MENDELIAN SEGREGATION. The way that recessive characters will segregate in the proportion of 1 to 3 in a second generation of a cross between a dominant and a recessive gene plant. *See* page 31.

MONOCOTYLEDON. Plants producing only a single seed-leaf on germination.

MONOECIOUS. Plants with unisexual flowers, but bearing both sexes on the same plant.

PETALOID. Resembling a petal.

PETIOLE. A leaf stalk.

PISTIL. The female part of a flower comprising the stigma, style and ovary.

POLYPLOID. Containing multiples of more than twice the haploid number.

RAY FLORET. The petaloid florets in the *Compositae*. A thistle for example is composed entirely of ray florets. Daisies contain an outer ring of ray florets and a centre of disk florets.

REMONTANT. Repeatedly bearing flowers in the course of a single season.

STAMEN. The male part of a flower consisting of a filament on which is the anther, which bears the pollen.

STAMINODE. A rudimentary stamen, that bears no pollen.

TEPAL. A sepal that resembles a petal.

TETRAPLOID. Containing four times the haploid number of chromosomes in each nucleus.

VARIETAS. A population of a species that is found in the wild and differs in some particular character from the typical form. Such varietates will breed true if self-pollinated or pollinated amongst themselves. The word is usually abbreviated var.

VILLOUS. Woolly or shaggy.

## NOTE ON REMONTANT IRISES

In comparatively recent years there has been a new race bred of bearded irises, which are classed as remontant. These bloom at the normal time and have a second blooming in the autumn. This remontancy seems to depend to a considerable degree upon climate. Some cvs will rebloom in California, which are single flowering further north and Australian growers report reblooming in a wide variety of cvs., which have not been reported as reblooming elsewhere.

However, the phenomenon is not entirely climatic and remontancy has been reported in a number of species. It may be a coincidence, but they are chiefly dwarf species with 48 chromosomes: *II aphylla, subbiflora* and *balkana. I. aphylla* has been used quite extensively in breeding, but the other two species rather sparingly. A leading American breeder of this race, Mr Edwin Rundlett, tells me that most of the remontant cvs in north-west U.S.A. seem to show traces of *variegata* and *pallida.*

This remontancy is genetic and is transferred to second and later generations. Our leading breeder, Mr Maurice Peach, has been successful using American cvs and a remontant strain of *I. aphylla.* Mr Rundlett points out that remontancy often occurs when sterile offsprings are produced from crossing 48 chromosome Tall Bearded Irises with the 40 chromosome *I. chamaeiris* and suggests that the energy that would be employed in producing seeds is transferred into producing a second crop of flowers. But though this might explain the phenomenon in some sterile hybrids, it will not explain it in fertile plants. In any case such sterile irises as 'germanica' and 'albicans' show no tendency to remontancy.

A possible explanation is that remontancy is due to a recessive gene, which emerges more frequently in some species than in others. If this is bred into a T.B. it will be inhibited, but if two T.B.s with a remontant species in their ancestry are crossed, some of the offspring will show this feature, and, if these are crossed, this remontancy will be perpetuated. This explanation, which, so far as I know is my own, sounds a little too glib to be true, but it does seem plausible. The effect of climate on remontancy is understandable, as many T.B.s will not flower at all in wet climates and it seems clear that a certain amount of summer heat is necessary to initiate flower production. One would expect more remontancy in districts where a hot summer followed by a wet autumn can be relied upon. In Britain many of the Oncobreds are unsatisfactory in their flower production, while in central and southern U.S.A. they are spectacular.

As a garden plant the remontant bearded iris seems to be one of the most hopeful developments of modern times. It is true that at the moment the colours of the remontants are less satisfactory than those of the single blooming cvs., but it is presumably only a matter of time before this disadvantage is overcome.

# Index